Iron Messiah

A Novel by

J. Schimschal

Iron Messiah is the third novel in the Darken Realm series of books and is the sequel to The Devil's Utopia and Ruins of America

Iron Messiah

Published by Fossil Ridge Books Inc.
P.O. Box 33218
Northglenn, CO. 80233

ISBN 10 0-9777327-2-x
ISBN 13 978-0-9777327-2-2

Published in the United States of America

Acknowledgements
This book is dedicated to Lily. May you always attain your dreams in life by following your heart.

About the Author

J. Schimschal is the author of the Darken Realm series of books. He lives in the western United States with his family. Additional information can be obtained by visiting www.darkenrealm.com.

About the Photographer

The wonderful cover art was captured by Michael and Nancy Bray. Michael and Nancy are talented photographers that live in Colorado.

Prologue

After fighting the very will of evil itself, Mineera, betrayer of the Dark Order has finally driven out the insane voices plaguing her fragile mind. Having survived the call of evil, she has finally severed the final tie with her former masters. Without the taint of darkness surrounding her, Mineera removed the oppressive shackles of her past existence.

As her new found freedom acts as a beacon of light for her, darkness is plotting the demise of her companions. Having tracked Nova 7 from the battered city state of Rasheed, Guillotine, hired assassin in the service of Queen Toil, has finally made contact with the rest of the mercenary team.

Hiding in the shadows, the vile assassin lies in wait, patiently watching and waiting for an opportune time to strike. Minutes pass as he watches his prey with sick fascination. Finally, an opening presents itself.

Feeling the sinister mutant stalking her friends, Mineera, lost in the tunnels underneath the ruined city, charges to rescue them from certain death…

Chapter 1
Whispers of the Ancients

The screen flickered and light erupted in the darkness of the eerie tunnel. The fan in the ancient laptop computer sprung to life, whirring with a dull hum which echoed within the secluded passage in the Iron Gate ruins. A moment passed, and the liquid crystal display began to produce a kaleidoscope of colors. Dull blue circles surrounded by a yellow background lit the darkness, flooding the tribal scholar in a wash of warm hues. The light emanating from the screen reflected in a pair of circular, wire-rim glasses. A pair of green eyes, hungry for lost knowledge, scanned the flickering screen with a comical, child-like gaze. Tani smiled as images appeared upon the screen of his most prized possession. Allowing his mind to settle, he prepared to explore a domain of thought and endless possibility. He extended his right hand, his stumpy fingers beginning to bang away at the keyboard. The keys clicked under his attention, responding to the mangled hand pressing down upon them. Smiling again, the tribal scholar began to explore a nearly forgotten world, a world filled with knowledge.

Trembling with elation, he submerged himself in the treasure clutched in his stumpy hand. A pile of worn, round, shining disks were his prize, and he was shuffling through them quickly. Tani's gaze frantically scanned the selection of ancient disks, eyes bulging eagerly behind his glasses. There was limitless possibility contained on the ancient disks. What did they hold? What lost knowledge was about to be discovered by the wily scholar?

Curiosity was Tani's bane. The tribal had scoured his

surroundings immediately after Jared and Banion left to patrol the tunnels around their makeshift camp, in the aftermath of Mineera's violent actions. It had been hours since her disappearance and still, there was no sign of her. The silent and foreboding network of passages had concealed any trace of her passing. Banion and Jared had been gone for nearly half an hour already. In their absence, Tani had explored the ruined tunnels on his own.

While sifting through the ruins of the ancients, the young scholar had encountered a collapsed music store. Ages ago, the earth had split apart and the merchandise had been channeled into the rift, collapsing into an enormous pile of lost knowledge. The collection of music was the combined creative fury of a now dead race. Tani, in his exploration, had discovered the forgotten treasure trove. With an exuberant lust to discover the lost secrets of the ancients, the young scholar had liberated several dozen of the disks from their plastic cases. Scampering back to the camp, Tani had to forcibly control himself, almost overcome by excitement over the unknown contents of the shiny disks. Glad that his companions were gone for the time being, he had resolved to load every one of the disks onto his laptop and explore the world of the ancients.

The disk drive opened on the side of the computer. Grinning wildly, Tani placed a disk on the shiny metal post and shut the computer drive. Drumming his fingers against his knee, the anxious youth was impatiently anticipating the lost knowledge contained on the ancient disk. What did the shiny treasures hold? With hundreds of such disks still littering a nearby passage, the possibilities seemed endless, especially to a wildly creative scholar.

It took a brief moment for the ancient laptop to read the information contained on the disk. A whirring noise erupted from the drive as the laser began to scan its surface. With a jolt of excitement, Tani hoped his computer could still read the ancient disks. Nearly holding his breath, he felt as if the anxiety might kill him before he got to find out.

"Come on!" he said in an irritated tone. The whirring of the disk drive seemed to be taunting the youth. Time passed at an agonizing rate. Each second seemed like an eternity. Suddenly, the old world data erupted from the computer with an unexpected result.

The speakers crackled and a faint hum began to roll from the computer. At first, Tani thought it was just his imagination.

Growing in intensity, the dull sound rose, and a series of low pitched whines emerged. Not sure what to think, he listened intently. Blinking several times, he let the sounds fill his mind and soul.

The haunting strains of a chorus of violins sounded in the darkness. Holding his breath, he heard the rhythm and was stunned. As the notes raised the hair on the back of his neck, the eerie tune awoke his powerful mind. Like gears grinding behind his green eyes, Tani's brain pondered the strange music, loving every note and beat. Smiling, the tribal scholar listened intently as the instrument whined, its ghostly sound echoing into the tunnels. As the violin's strains rose in intensity, a host of other instruments joined the lonely sound. The light pounding of drums could be heard, adding to the rhythm. Horns, cymbals, and other unknown musical instruments rolled in and out, filling the abandoned ruins with a haunting symphony of sound.

The music was so intricate and complicated; the young scholar's mind was set ablaze. The tones and sound were so perfectly matched, Tani marveled as the symphony continued to unfurl from his laptop computer. It was as if some ancient god had forged the song from the very depths of its heart, taking part of its soul to permanently harness pure emotion in the form of music. Shaking his head, the tribal scholar smiled and felt very lucky.

Listening to the music echoing, Tani was in awe. He looked at the case of the aged disk and saw the name and image of a classical composer, a person who had not breathed the air nor walked the earth in over one thousand years. In that moment, as the thoughts and emotions of a composer dead for a millennium entered the tribal's mind, Tani felt small, very small and insignificant. Marveling at the thought that the disk had survived all of the chaos of the apocalypse and the passage of time, he saw it as a wonder unmatched. The realization that a person could find such a treasure amidst the debris and ruins of old was a staggering thought to Tani. If this one small disk contained such wonderful music, what did all the other disks contain? Flipping through the pile of shiny disks, Tani smiled; such knowledge was power. Letting the sounds fill his mind and soul with their haunting melody, Tani stood up, still clutching the pile of disks.

As he stood in the dim light of the tunnel, the scholar's eyes moved toward a yellow glow just south of his location. Lava from the bowels of the earth coursed through the chamber beyond, lighting the darkness in an eerie yellow glow. The scholar blinked several times as he stared at the doorway, the same doorway that, just a few hours ago, had set the stage for a wild, mad scene. In that frantic flurry of events, Mineera had tried to kill young Jared in a fit of rage and madness.

The desperate attack, and the subsequent loss of the tormented psychic in the bowels of the underworld ruins, was still stinging his mind. *Somewhere* in the darkness was Mineera, a woman driven to the edge by insane visions and a never-ending host of maddening voices, pulling at the fragile core of her being. Mineera was still in the ruins and suddenly, Tani felt very foolish.

Listening to the music pouring from the computer, he felt fear rise at the edge of his soul. If she was out there, somewhere, the music would surely bring her back like a trail of breadcrumbs through a dark forest. Banion and Jared were still patrolling the tunnels, and this knowledge put the young scholar further on edge. If Mineera did return, Tani would be essentially defenseless. Her potent psychic abilities could easily overwhelm him, and he would be ill prepared for an attack. Banion was convinced that Mineera would return from the darkness, healed and strong, but Tani was not so certain. After all, she had tried to kill Jared. If she did return from the dark tunnels, would Mineera be gripped by evil and madness, or cured from her insanity?

With a frantic motion, Tani rushed to his computer and stopped the music. The sound ceased and the dark tunnel was oddly silent. Looking around with a sigh of relief, Tani viewed the crumbling tunnels around him. Cracked concrete beams had splintered, littering the floor. Water trickled down and pooled near the entrance of the tunnel. His hands clutching each other tightly, he sighed, and was suddenly overcome by a strange feeling. In the dim light of the tunnel, something felt odd; something didn't feel right. Uneasy, he whirled around as if *something* was staring at him. Shuddering with a chill in his heart, Tani felt as if he was being watched.

Whirling around again in the opposite direction, the tribal scholar peered out into the darkness. The icy blackness responded only with uncertainty. Whatever lurked in the shadows was protected by the darkness, caressed by its loving silence and cloying seclusion.   The hair on the tribal's neck stood on end again. Shivering, he peered around him with frantic glances, still feeling as if something was watching him.

"Just my imagination…" Tani said with an uneasy look in his eye. Trying to shrug off the uneasiness, the tribal youth looked towards his gear, the backpack which contained all of his weapons and means of defense. With an expression of shock, Tani's eyes scanned the ruined tunnel, darting back and forth, increasingly desperate. Something was missing…

The uneasiness quickly turned to all-out panic. The tribal's backpack was *gone*. All of his weapons — *gone*. Scrambling around in the shadows, Tani searched frantically for his belongings. He blinked several times, trying to rationalize the loss. Not wanting to believe the eerie occurrence, the young scholar tried to formulate a theory that he had simply misplaced his backpack. The thought faded quickly as the nearby rubble of concrete crackled under the weight of *something* moving within the shadowy darkness.

"Jared?" Tani cried out in a frightened, high pitched tone.

Nothing and no one responded.

"Banion? Is that you?"

More silence.

"Come on, guys, this isn't funny!" the tribal whined, feeling the hair on his arms stand on end.

Something was still watching him, hiding in the ruins.

"Mineera…" His voice dissipated into a feeble whisper. Losing his ability to speak, the young scholar quivered.

All of his weapons were in his backpack. His submachine gun, explosives, and even Tani's hunting knife were gone. Plain and simple, the tribal scholar was defenseless. His dread mounting, he spun around quickly again, scanning the ruins with an uneasy gaze. With his vision obscured by the shadow, his other senses were on overload, trying to detect the source of his increasing apprehension.

The sound of metal sliding free echoed through the ruins. A dagger, a jagged knife, had just emerged from its sheath. Cold steel was hungering for warm blood. Edging back, Tani moved away from the sound. *Whatever* was out there, it had just pulled a knife.

"Banion!" Tani screamed in panic. His mind was wild with fright. The young scholar wanted nothing more than to have the reckless gunfighter charge forth and save him from the lurking menace. But there would be no help in this struggle.

Something launched forward, at Tani's back. The rush of movement made the scholar jump. The feeling of sharp metal piercing his flesh throbbed through his nerve endings. Dodging away from the flurry of movement, spinning, and shrinking from the attack, Tani felt pain erupt in his arm. Stunned, he clutched the wound as warm blood flowed freely down his left arm.

As slick blood dripped from the wound, another flurry of movement rushed forward. Instinctively, wanting to defend himself, Tani flailed his arms toward the onslaught of the unknown ambusher. The sharp knife slashed him again, tearing into skin, shredding his arm. The blade struck bone. As he reeled back, the tip of the knife was imbedded in his forearm. A piercing flash of pain emerged from the wound like a spike of fire. Yelping, he flung the attacker away from him.

As his mind spun desperately, wild thoughts were his only companion. Was it her? Was it Mineera, returned from the darkness to continue her fit of madness?

Stumbling backwards, Tani toppled to the ground, his wire-rimmed glasses falling from his face, landing somewhere in the darkness. The lights went out. Poor eyesight is often a curse, but in the midst of battle, it can be downright deadly. His field of vision was an obscured blur. As blood dripped down his arms, Tani screamed in terror, "No!"

Blinking, flailing about like a dying fish, Tani strained to catch sight of his attacker.

"Oh, yes, oh, yes indeed!" A low hiss sliced through the darkness. Gray and white fur glimmered in the dim light. The mutant opossum crouched in an aggressive stance, bloody knife in hand. His black eyes stared at the fallen Tani as his tail swished back and forth, anticipating the kill. "Mistress Marion will pay much for your rotten skull!"

Leaping forward without a hint of mercy, Guillotine landed on top of the downed Tani, slashing away at the defenseless scholar. Screaming in terror, the young tribal flailed about. The weight of the mutant assassin was now upon him. The opossum's sharp, bristly fur pressed against him, and the stench of Guillotine's musky body pushed Tani deeper into a state of terror. Guillotine had only one objective: to kill young Tani and cash in on a hefty bounty. The scholar was alone and knew his chance of survival was slim. Other than Banion, the crazed mutant opossum was the most dangerous thing in the mercenary world.

Desperate to stay alive, Tani grasped Guillotine's furry arms, keeping the sharp knife from piercing his flesh once more. As he flexed his arms, struggling for his very survival, his wounds bled fiercely. The strength of the mutant under whose weight he quivered was immense. Slowly but steadily, the knife edged forward, toward Tani's throat. He shook with effort, his strength fading, and the fear of death all too real. The tribal knew his life was about to end, alone, in the dark tunnels beneath the ruined city.

"Can you feel the darkness, little one?" Guillotine rasped, watching the point of his blade connect with the skin on Tani's throat. The tip of the knife dug in like a needle, breaking the skin; a single drop of blood emerged from the small wound. "Give up, little one, and I will spare your suffering. Give up and your death will be swift!"

As he felt the knife's edge digging into his throat, tears rolled down Tani's face. Frustration filled him, draining him of his will to survive. Death was drawing close, and the scholar's mind began to flash with images from his life. Screaming in the ruins, Tani shook, nearly sobbing, holding back the weight of the mutant atop him with every shred of determination in his body. The knife point began to sink deeper into his throat. His strength failing, Tani of Scarskin prepared to meet his ancestors.

Chapter 2
Redemption

A sickly feeling of fear coursed through her. Someone near was fighting for their very life and was close to death. The emotion of sheer terror was strong, and Mineera tried to focus while working her way through the labyrinth of abandoned tunnels far beneath the once-mighty city.

"Do not fret, your faith is now absolute." A whisper rolled through her mind. The voice was familiar and soothing. Knowing that the being giving her a spiritual push was a thing born of light, Mineera took solace in its calm presence. *"Your strength has finally liberated you from the darkness. Feel the freedom you have earned, prophet of Gogoli..."*

A powerful feeling surged into Mineera. A blast of pure warmth rushed into her body, working its way through her being like a tingle of electricity. It was a calm, inspiring sensation that almost made her feel as if she were floating. No longer was the taint of evil a tangible thing, tearing at her soul. The whispers in her mind were gone, a fading memory relegated only to the frightening guise of dreams. No longer did the devil hold sway over her fragile sanity. At last, the darkness had been driven away. A warm surge of power was now inside her, a spiritual power far beyond anything she could have ever imagined. Mineera was finally free from evil and walking a path of righteousness, rather than traveling a tainted road to ruin.

"The throes of evil have abated. What you feel inside you is a belief and power born from heaven itself. Your faith is your

primary power, but the agents of the enemy must be quelled by a higher force." The strange being spoke as Mineera continued to rush along the ancient tunnel with flashlight in hand. The search was becoming frantic as she sensed Tani's familiar presence nearby. The scholar's normally whimsical aura was marred by black thoughts of despair and terror. It was an unsettling feeling for Mineera. In her heart, she knew that Tani was in great danger. As she pressed forward with her mind, the sickly feeling increased tenfold. Mineera was now certain that the fear she was feeling was emanating from the tribal scholar.

"Trust in your senses. If needed, you can channel your energy into a focused burst of power. No enemy born of darkness can stand against such power. Use it wisely, Mineera of Gogoli. For while your current strength is great, it can be used regardless of intent, a savage power without recourse or morality. You must control yourself and your new power. Without a strong resolve, even the most righteous can fall to ruin, ripped by corruption. In times of indecision, trust in your heart, not in any voice raging in your mind." With these parting words, the spirit began to dissipate, falling away like a whisper into the shadows. In the blink of an eye, the presence was gone, leaving Mineera to forge her own path and destiny.

Left to her own thoughts, she could sense herself drawing near to the scene of conflict.

"Hold on, Tani..." she whispered, running down a dark passage hewn from the bowels of the earth.

The eerie glow of yellow flame loomed ahead. Feeling the familiar sparkle in her mind, Mineera could see the edge of the room leading to the fiery pit of lava, the same room in which she had ambushed Jared only a few hours ago in a fit of rage.

Entering the room, Mineera gasped in horror at what she saw. Guillotine, vile mercenary and minion of the sinister Marion Toil, was on top of Tani, trying to force a bloody knife into his throat. The full weight of the sinister creature was pressing down with all its might. The trembling Tani was trying desperately to keep the sharp blade of the knife from pressing deeper into the flesh of his throat.

A chill rolled down her spine. It moved to the bottom of her feet and spread out into the earth. Tani would be dead if she did not act quickly. The feeling of urgency pushed through her like instinct. She bowed her head, her eyes zooming in upon Guillotine as he bore down for the kill. A calm came over Mineera. A pulse of pure energy rolled from her mind, snaking down her arm with a dull white glow. Extending her hand forward, she leveled her fingers towards Guillotine. She shook from head to toe, the energy inside her raw and powerful. Concentrating, Mineera focused on attacking Guillotine with her newfound powers.

It felt strange at first, sensing the power of heaven being focused in her body. As she lent her thoughts towards serenity, the power rose in a heaving wave, a surge of pure energy. Shaking, feeling the new powers within her, she fought to channel the energy without losing her grip on the conflict. A pulse of warmth exploded around her wrist, and her powers erupted in a violent display.

The blast of energy sped forward in a blinding beam of light, as bright as the sun itself. Breaching the void of darkness, the holy blast of light lit the room like a bolt of lightning. The darkness was torn asunder as the blazing beam of energy struck Guillotine in the chest. The searing holy flame was so powerful, it lifted the sinister mercenary off the fallen Tani and hurled him against a nearby wall. Slamming into the concrete, Guillotine gasped as the holy fire burnt his flesh. Sliding down the wall, the mutant opossum yelled in agony; the blast was so potent that part of the steel breastplate that protected the vile mercenary had melted. The smell of burnt flesh and fur filled the room. With smoke rising off his singed body, Guillotine made a feeble attempt to stand.

Still wracked with pain, the sinister mutant pulled a handgun. He hissed, shaking with rage, and opened fire on Mineera. A volley of gunfire roared relentlessly toward the psychic.

Already anticipating his next move, Mineera was ready for the hasty attack. Undaunted, she formed a barrier of energy around herself. A psychic shield flared instantly around her body, keeping her safe from harm, protecting her from the gunfire. The holy white glow from the mystical barrier was so bright in the darkness of the earth that Guillotine was blinded by its radiance. The bullets rammed into the barrier of psychic energy and were harmlessly deflected. Holding his clawed hands in front of his face to shield his eyes from the burning light, Guillotine had to shake off the bright spots that flooded his vision. It was as if his retinas were burned by the glowing shield of energy. Knowing full well that Mineera was a psychic warrior from the highest ranks of the Reaper Kai order, Guillotine felt that he was severely outmatched. He was right; she was now even more powerful than ever before, the embodiment of justice on earth.

Knowing he was seconds away from a fiery death, Guillotine squinted in the bright light, buffeted by the conflicting forces of self preservation and rage. With an angry taunt, the mercenary hissed once more and screamed at Mineera. "Foul bitch, your deeds this day will not go unpunished!"

"Nor will yours. Flee now or face me!" she responded in an ominous, commanding tone. She gazed at Guillotine with a dominating presence locked behind her soft blue eyes. Still squinting, he looked into those eyes and knew she meant business. Abandoning his prey, Guillotine turned away and looked for the nearest route of escape.

Retreating, the sinister Guillotine leapt away into the darkness of the tunnels. As he fled the scene, the smell of burnt fur and flesh went with him. Dazed with pain and nearly blind, the vile assassin was driven back, forced into the darkness of the ruins by the righteous power of the psychic warrior. Mineera let him go, wishing only to focus on the wounded Tani.

The threat of combat was now over. As she allowed her senses to return to normal, fatigue stung at her mind. The toll of her newfound powers had drained her of spiritual energy. Feeling a dizzy wave wash over her, she staggered over to the fallen scholar, her blue eyes blinking quickly in an exhausted flurry.

With a flash of fear in her heart, she stared at her fallen companion. The tribal did not stir. Even as she rushed toward him, Mineera's hope was beginning to fade. Lying prone, covered in his own blood, Tani did not move a muscle. She shook with fear and frustration, collapsing to the ground. Holding her breath, Mineera stretched out her hand, moving it toward Tani. Inch by inch, her hand drew closer, as Mineera clung to the fragile hope that the young tribal had survived the sinister ambush.

Chapter 3
Quelling the Fear

With a trembling hand, Mineera reached forward to touch the fallen Tani. As she grasped his shoulder, she held her breath. Unable to tell if he was alive or dead, she prepared herself for the worst. In the course of the battle with Guillotine, Mineera had expended much spiritual energy and was unable to use her powers to probe him for signs of life. Relying on traditional means instead, she touched him again, pushing his shoulder with a reserved, tense motion.

The result of her touch was a sudden explosion of movement. With a jolt of panic, the wounded scholar lurched back, screeching out an alarmed yelp. The fallen tribal was definitely alive. He crawled backwards across the rubble, shielding himself from Mineera by throwing one of his bloody arms in front of his face, trying feebly to shield himself from harm with his outstretched hand. Horrified by the ambush, the scholar was barely rational. He was still unable to see after losing his glasses, and was gripped with fear, not knowing that the sinister Guillotine had been driven away.

"Tani!" Mineera cried out in concern, holding her hands in the air in a gesture of submission. Still dazed from the attack, Tani blinked several times as the sound of her voice rang out in his ears. In his terrified state, Tani still considered Mineera a foe instead of a savior.

"Stay back!" he screeched again, his voice cracking in fright.

"Be still," she responded. Her speech was soothing, but the young scholar was still out of control. After all, he had just survived an aggressive attack and was still rattled by Mineera's recent betrayal. Trust was not in the tribal's vocabulary at the moment.

His sanity was slowly returning. Breathing erratically, he tried to get his emotions under control, frantically attempting to wipe away the tears from his face with hands which were still bloody. Streaks of blood mingled with his tears, and he felt humiliated. Tani wanted nothing more than to disappear into the background and hide from the entire ordeal. Turning away from her, he whined again, "Just leave me alone!"

"I can't do that," she said, her voice compassionate. "I just can't do that." Her somber blue eyes fixed upon him, she tried to coax him into trusting her once more.

"I can take care of myself!" Although he was doing his best to sound heroic, the blood-covered scholar was not convincing. His words were in complete contrast to his actual emotions. In reality, Tani needed someone to rescue him from the fear still coursing through him.

"You are wounded. Let me help you." Mineera began to scan their dim surroundings, groping around on the ground near Tani in search of his spectacles. Quickly, she found what she was looking for. "Take your glasses." Her voice was soothing, and she handed the spectacles to the frightened scholar with a slow gesture to keep from frightening him further. His green eyes tried to focus on the glasses in her hand. With a trembling lip, he reached out and grabbed them from her. Mineera had shown him kindness when he had been trapped in nearly impossible circumstances. If she had intended to cause his demise, she would have already killed him, especially in his weakened state. Even though she had betrayed Nova 7 in a fit of madness, the uncertainty of her mental condition was a welcome alternative to the vicious aggression of Guillotine. Focusing on her gentle motions and soothing words, the young scholar was already beginning to come out of his shell and regain his trust in her.

After putting his glasses back on, Tani sat up and stared at Mineera silently. He clutched at the wounds on his arms and throat pitifully, avoiding direct eye contact with her. He would look at her with quick, darting glances, then avert his gaze as if ashamed of his condition. His reservations began to abate as he looked at her, breathing softly beside him. Returning his focus to his wounds,

Tani looked at himself in disgust.

Blood covered his fingers. Deep gashes had been cut in both of his forearms. A dull throbbing hurt was focused at the center of his throat. Falling into an uneasy calm, he stared at Mineera with soft eyes, almost remorseful. It was as if he wanted to apologize for his condition. Feeling his pain, Mineera shook her head as if to tell him that it was all right, everything would be fine.

"You didn't do anything wrong," she blurted out, still feeling his raw emotion clawing at the edge of her senses.

Tani did not respond. Instead, he looked directly at her, peering intently into Mineera's peaceful, consoling eyes. As the seconds passed, the silent stare turned to a whimper. His lip trembling, tears began to roll down his cheeks. Tani began to sob uncontrollably as he surveyed his bloodstained wounds. The tears rolled down his cheeks in hurried bursts of fear as his whole body shook in frustration. Still staring at Mineera with sheer terror in his eyes, Tani was a ruined mess.

Moving forward, Mineera furrowed her brow and extended her dark skinned hand. Tani watched it slowly, secretly wanting to be consoled with sympathy, to have someone protect him from his failed courage. Her hand touched his shoulder with a gentle pat. Responding to her touch, Tani shook his head back and forth quickly, still sobbing. Like a frightened rabbit, he shook and twitched, shifting about nervously.

"I was so scared..." the scholar whispered, snot streaking around his nostrils. Still frightened and covered in his own blood, Tani didn't even care about his appearance.

"You're all right now," his companion reassured him.

Nodding vigorously, trying to fight back the tears, he agreed silently. "Are you all right now?" Tani shot back with a slight tremor of fear. When last they met, the crazed psychic had tried to end Jared's life. It was a deed not easily forgotten.

A moment of silence hovered between them. Their eyes were locked, each still unsure of the other. Finally, she replied with a smile, "The battle waged within my soul is over. I have claimed victory over the dark voices that have plagued my heart."

Looking at her, Tani tried to convince himself that the struggle within her soul had truly ended. The tremor and frantic look in her eyes were gone. No longer did she twitch and turn her head about like someone being plagued by a swarm of bees. All in all, she seemed different and calm, the look in her eyes almost serene. The change in Mineera's personality, from a tortured, crazed person to a placid, focused one, seemed to be genuine. Feeling that she could be trusted once more, Tani began to speak of the rest of Nova 7.

"Banion and Jared are out on patrol," he said in a matter-of-fact tone, probing the wounds on his throat with his fingers. Guillotine's knife had dug in deeply, but had not pierced any vital arteries. With feeble motions, Tani smeared his own blood about in a sickening display. Seeing his concern, Mineera pulled his arms down.

"Be still," she commanded the youth. Uneasy but less suspicious of her, Tani dropped his hands. "Let me have a look at your throat." Slowly, she took out her knife, not wanting to startle Tani. Mineera cut a swath from her blue robes and fashioned a crude bandage. Using her canteen, she washed much of the blood away from the area around the wound. Securing the bandage around his neck, Tani felt more at ease.

The sobbing had ended, turning to a whimper with heavy, labored breathing; the tribal was still obviously stressed from the horrific event he had just experienced and survived. The flow of blood around his throat slowed drastically as it clotted under the pressure of Mineera's makeshift bandage.

"Thanks..." Tani whispered, his eyes distant. The tribal almost appeared to be in shock, and with the amount of blood he had lost, it wouldn't be surprising. He shivered. "I'm so cold."

"You have lost a lot of blood. Just be still."

"I was so scared. I almost gave up. Never have I felt such terror. I wanted to surrender so the pain and fear would end. I almost let Guillotine kill me just to free myself from the panic and pain," the tribal rambled, still shivering from the blood loss and shock brought on by his wounds.

"I'm glad you didn't surrender."

"Me too," he whispered back. Tani stared at Mineera through his wire-rim glasses. With a twist of his lip, he smiled faintly at her.

Still mindful of her goal, Mineera pulled his arms forward and surveyed his defensive wounds. Blood had already clotted on the horrid gashes, but they still needed attention. Cutting two more swaths from her robes, Mineera bandaged and cleaned his wounded arms.

"What if he comes back?" Tani suddenly blurted out, trying to rise to his feet. As he tried to stand, Mineera resisted him, pulling him down.

"He's not coming back, not today." She spoke in a confident tone. "Lean back, take some rest. You have been through quite a lot."

Heeding her advice and weak from blood loss, the tribal leaned back, and Mineera drew near him, soaking in his fear and pain with her mind as her psychic powers probed his being. Wanting nothing more than to wash this pain away, Mineera suddenly felt guilty. The peaceful tribals and Banion had treated her with compassion, and she had betrayed them all in a fit of madness. Knowing that she had crossed the line, Mineera felt that Banion and especially Jared would be reluctant to welcome her back. Tani was eager to forget her betrayal, feeling she had been vindicated by saving him from the vicious Guillotine. But all in all, she was facing an uneasy homecoming.

Tani, still dizzy, looked quietly at the distress on her face. The scholar wasn't a psychic, but could tell the cause of Mineera's anxiety nonetheless. "We are all in this together. Each and every one of us," Tani reassured her. Mineera was not convinced.

"I betrayed you all. That is something I can never forgive." Her voice was distant and she sighed in despair.

"It's not up to you what is forgiven." The statement was filled with wisdom, so much wisdom that Mineera took note and felt comforted. Another calm stretched between them as the wounded tribal considered her softly, lost in thought.

The tribal's green eyes stared into her own. Over the last months, she had had a wild, feral look hidden beneath her striking blue eyes. The primal, haunted look was now gone. Only a serene peace filled her blue eyes. She had been cured of madness once and for all, and the tribal could tell Mineera had truly come back from the darkness. Feeling the stress of the assault linger, the scholar tried to console his savior.

"For the first time in months, I see you, Mineera, the real you. If I can see it, they'll see it. The road we are on has bruised us all. None of us are immune to this horror…" Tani touched the bandage on his throat, still feeling the bloody knife pressing into his flesh. With a shudder, he pulled his hand away quickly. "You betrayed us, yes. But if it were not for you…" New tears welled in Tani's eyes. "… I would be dead."

The two companions stared at each other, one a creature of science, the other of the spirit. They were unlikely companions, bitter rivals in thought and mind, but they were companions nonetheless, two souls brought together by the same goal: liberating the world from the scent of darkness still spreading across the land like a sickening plague.

One would find salvation in the force of knowledge, while the other would find strength in the mysteries of the soul. In the darkness of the tunnel, Mineera and Tani came to a silent truce. Tani would still deny the spirit world and Mineera would scoff at technology. It was a stalemate that both were willing to accept. Turning aside old differences, Tani stretched out his stumpy fingers. Mineera looked back and placed her hand on his. They gripped each other's hands and held them still for a brief moment. Then they smiled at each other, their hands falling away.

Suddenly, a flash of mischief rolled through Tani's light green eyes. The look grew and Mineera sensed a rebellious wash of emotion sweep through his mind. With a weak whisper, the tribal taunted her. "There is no God."

Mineera was stunned by his bravado, but shot back immediately, taunting the youth in return. "There is a God, only you lack the wisdom to open your mind, because you are too busy fidgeting with gadgets and that damn computer of yours."

A comical look illuminated both their faces. Each started to giggle, and soon erupted into a burst of laughter; the two companions felt a peace emerge between them. A truce had been struck with a smile and deep respect for one another. They would never see eye-to-eye on religion and science, but it didn't matter. They now shared a link that could never be broken, a tense emotional event that tied them together with a bond of trust.

Chapter 4
Sadistic Vanity

Still trembling, Brother Feral, sinister warlord of the Reaper Kai, gripped his hands tightly. The momentary loss of control had overwhelmed his senses. An empty, hollow sense of loss filled him as he surveyed his chambers. Rising from his dining table, he twitched violently as dark thoughts mastered his soul. Lost in contemplation, the Reaper Kai general slowly walked to the corner of the room.

Coming before a great mirror in his chamber, Feral admired the perfect body reflecting back at him. He smiled in response to the obliging bulge of his muscles as he flexed them, his own eyes hungrily watching his reflection.

"So beautiful..." he whispered, turning from side to side so he could get a better look at himself. Marveling at his finely tan skin, a dark smile graced his lips. He fell into an almost trance-like state, dedicating several long moments to viewing himself. While he lingered over every inch of his well developed body, the strange fantasy turned to anger as Feral caught sight of a stain upon his wrists and forearms.

Part of his evening meal was littering his garb. Seeing the filth upon him filled Feral with a deep rage once more. His perfect form could not be marred by any blemish, let alone refuse, especially as it had been deposited upon him by something inferior, a feeble servant. During Feral's evening meal, the servant had stumbled, causing his nightly meal to land upon him. It was a simple accident, but Feral was prone to fits of uncontrollable rage.

After seeing his perfect body marred by the accident, he lost control. Within a mere minute, the Reaper Kai general had beaten the servant to death. An innocent person had died simply because of a tiny mistake. As he recalled the event, another spike of anger rocked him. Exploding once more, he strode over to his table and began to scream at the lifeless servant.

"How dare you! Look at what you did to me! I am stained! Stained by a filthy wretch like yourself!" he wailed at the lifeless body next to his dining table. The lifeless body did not respond to his taunts or his wild screams of anger.

Unable to defuse his rage, Feral opened the door to his chamber and began to shout, "Get in here now!"

Two female attendants, slaves of the Reaper Kai, moved forward slowly, averting their gaze so as not to anger the sinister general further.

"Yes, my lord?" one of the slaves replied in a submissive tone.

"Get in my chamber and clean it!" he yelled in rage.

Following his orders, the slaves rushed inside and began to attend to the room. Though such incidents weren't a common occurrence, this wasn't the first time that Feral had killed one of his servants during a psychotic burst of animalistic rage. Finding the body on the floor was frightening to the slaves, but paled in comparison to disobeying Feral's orders. Driven by a need to follow his command quickly, the two maids put aside their fears and removed the body from his dining area. After pulling the corpse out of the room, they quickly removed the stained tablecloth from his table and brought a fresh one. The clean-up activities were hasty and had been done in such a manner that Feral was not disturbed. Closing the door to his chamber, the maids disappeared into the palace, towing the body of his latest victim between them.

Using his napkin, Feral wiped the food from his robes. He was once again at peace. The vain priest returned to the mirror, beginning to survey his perfect body once more. No longer did food from the unfortunate accident cover his arms. The imperfection had been washed away, and he was now free to study himself again. Flexing his muscles, he posed and admired himself for another few moments.

Brother Feral was a monument to personal perfection.

Having studied the dark arts for years, the sinister priest had learned to commune with demons who were, like himself, obsessed with carnal pleasures and the pursuit of vanity. Following the ways of these demons and harnessing his natural, aggressive rage, he quickly rose through the ranks of the Reaper Kai order. His psychic abilities were more powerful than most of his counterparts, but his combat prowess coupled with his psychic abilities was what made him downright dangerous. In the entire Reaper Kai order, there was not a single more dangerous foe in close combat. Having studied almost every form of weapon, Feral could cleave and rend anything in his path as he saw fit. In most circumstances, he could destroy dozens of enemy troops in a flurry of psychotic attacks, driven by the primal rage simmering inside of him. This devastating combat ability, combined with his unusual psychic abilities, made Feral a potent warrior.

During his study of the dark arts and worship of demons, Feral had learned how to channel his spiritual energy into a focused burst of light. Enhanced by years of training, this ability to project light had given Feral the uncanny capability to bend light around himself long enough to vanish from sight. In the chaotic flurry of battle, this power was extremely potent. As the enemies of the Reaper Kai observed his ability to shift light, he seemed to be teleporting himself around the battlefield. To the free people, Feral was known by the name *Butcher Wraith*.

While his ability to disappear in battle by bending light around himself was a powerful gift, the most potent of his abilities was related to the perfection of his body. Obsessed with his appearance, gripped by fanatical vanity, his worship of demonic entities had gained him the favor of several powerful spirits. These spirits, possessed by an equally intense vanity, funneled supernatural energy into their beloved brother Feral. If he was marked or damaged in battle, the demons would grant him a burst of supernatural energy strong enough to heal the damage. In a sense, Feral's pursuit of vanity allowed him the privilege of never being scarred or deformed by battle wounds. If he was injured, his dark prayers would undo the damage in a mere day or two. This ability made it nearly impossible for the priest to perish, unless his body was torn by mortal wounds. Combining all of his talents together into one package made him a powerful warrior, but still paled in comparison to his ability to lead an army.

Where Sister Nightshade was reserved in combat, preferring to focus on tactics rather than brute strength, Brother Feral was just the opposite. His style of military command was based on aggressive, rapid attacks intended to demoralize the enemy into fleeing the battlefield; he then slaughtered his foes as they retreated. Feral was often seen spearheading assaults, leading powerful ranks of Goat Minions into the most heavily fortified of enemy positions. As he spread chaos amongst his enemies, preventing them from reacting quickly, most of Feral's foes found that they were overrun quickly and unable to counterattack effectively.

Though this wild style was often successful, the casualties inflicted upon the Reaper Kai forces were horrendous. Feral's forces would often suffer hundreds if not thousands of casualties within a few minutes of open, heated battle. This style of battle was reckless in the eyes of most Reaper Kai, but it didn't matter to Brother Feral; he had the greatest number of victories in military engagements. Victory was the ultimate success, despite high casualties, in Feral's eyes; cost was nothing but a number.

Still looking at his perfect body, Feral grew restless. Satisfied for the moment, he left the tall mirror and walked onto the balcony of his room. Cold night air had washed over the battered city of Rasheed. In the deepening night, the city below the walls of the palace was strangely silent. The oppressed people, the last remnants of Rasheed, had been utterly crushed into submission, without any hope of living out a peaceful life under the iron grip of the Reaper Kai.

Though the city below was devoid of life, it didn't bother the seasoned warlord of the Reaper Kai. Crushing empires under military action was what he enjoyed most when not in pursuit of carnal pleasures. Seeing the ghostly ruins below, he smiled. As he stared at the rubble of Rasheed, his thoughts began to focus on his next objective.

While Sister Nightshade had been given the honor of crushing the Steel Crag Mining Guild, a task that Feral coveted greatly, he remained undaunted. Father Vertigo had seen fit to allow Brother Feral the honor of crushing the second most powerful empire in the entire Darken Realm. His next objective was to lead an enormous army of Biogtechs and Reaper Kai into the northlands, with the intent of obliterating the Mord Tech empire. Already, enemy agents, under the control of Queen Toil, had infiltrated almost every level of the Mord Tech empire. When ordered to do so, these spies, assassins, and saboteurs would mobilize and shatter the Mord Tech infrastructure and command group. With the core of the empire in shambles, Feral and his host of soldiers would sweep in and secure victory over one of the great northern empires.

With wild thoughts of conquest filling his senses, his blood was set afire once more. Whether pursuing perverse obsessions of vanity and pleasure or crushing his opponents with his violent battle prowess, Feral was in the most gratifying position for one of his kind. Smiling with dark thoughts, Feral prepared for the rising of the sun. For when the light broke across the land, he would march out of Rasheed at the head of over ten thousand soldiers and war machines, with the intent of inflicting brutal slaughter and acts of fanatical hate against his enemies. The drums of war were beginning to rattle once more, and the target of the next attack was Mord Tech's capital city, Markov.

Chapter 5
Seeds of Victory

Crouching, the concealed gunman let the crosshairs of his mighty assault rifle come to rest on the chest of a blue robed woman near the back of the room.   He sized up the situation from his vantage point amongst the debris, surrounded by rubble, with light a scant companion.   Breathing softly he trained the weapon upon the woman's form, its crosshairs moving back and forth in sweeping motions across her body.   Drifting from her chest to head, then back again, the gun sights stalled as if the concealed gunman was having second thoughts.   His trigger finger quivered.   Should he kill her?   It would be so easy...   Just a squeeze of the trigger, and it would all be over.   Snapping out of the forbidden daydream, he searched the chamber for other people.

Using his weapon as his eyes, the concealed gunman caught sight of a young tribal, horribly wounded, covered in blood but still alive.   The lurking predator viewed the young tribal with concern.   His wounds seemed severe, life threatening, but the tribal was conscious and sitting next to the blue robed woman.   It didn't appear to the gunman that the blue-robed woman was the cause of the tribal's injuries.   If she had been responsible, the wounded tribal would not be sitting peacefully next to her.

The concealed gunman communicated with his companion using a series of hand gestures.   A stealthy young warrior moved from the darkness with a submachine gun in hand.   Aiming it at the woman garbed in blue, the youth considered her with a mixture of emotions.   As he eyed the wounded tribal next to her, he was filled with indecision and rage.

"What the hell happened to Tani?" Banion said, attempting to clear his clouded thoughts. It was apparent Tani had been wounded severely, but he was still lucid, still conscious. "He seems torn up, but he isn't afraid of *her*. It doesn't make any sense. Looks like we missed one hell of a show."

"I'm not sure what happened, either. If she did attack him, Mineera would have killed him. There is no way Tani could have survived." The tribal warrior shook his head in confusion. "Yeah, we did miss one hell of a show."

"What do we do now? If she loses control again, she could kill us all, compromise our mission. If we let her live, we're running a huge risk." Banion spoke in a whisper, training his weapon upon her chest once more.

"I'm not sure we should trust her, Banion. You saw what she tried to do to me. Is traveling with the enemy a sound decision? Can we put our faith in her?" Jared asked, training his submachine gun on Mineera as well.

With a sigh, Banion knew the decision was up to him. Sure, she had lost control and almost killed Jared, but there was something magical about her ability to stand against the will of evil on her own. Mineera was a maverick, a stout warrior who had been spiritually tormented, and yet had withstood the agony for many months as voices plagued her fragile mind.

Despite himself, Banion had softened somewhat over the past months of travel. His potent anger had turned from a roaring volcano to a simmering forest fire. The anger was still there, but it was in check, for the moment. His persona had changed, allowing him to move beyond blind reckless vendetta into a more evolved, obsessive sense of vengeance. Sure, she was dangerous, but he *needed* Mineera to finish the insane crusade. She had great skill and deep down, he knew that Mineera was different: not an ordinary, murderous Reaper Kai, but someone who had come back from the edge. Burying his deliberations, he made a hasty decision about Mineera's betrayal.

Banion sighed. "I say we pay them a visit," he said slowly. "See what this is all about. Stay cautious and keep her covered. If she flips again, do not hesitate; she was one of the Reaper Kai's most powerful." With a grim look on his face, he added, "I only hope the battle she was waging within her own soul is over."

Jared nodded back in agreement, grinding his teeth. Still hot under the collar about the ambush perpetrated by Mineera, the tribal warrior was uneasy about how he would react when they confronted her. His expression reserved, he prepared to follow Banion into the passage. As they approached, Jared's cheeks flushed with irritation. He felt a hostile rage begin to well up in his soul. Agitated, the young warrior flexed his grip on the submachine gun; Jared would not be caught unprepared again.

Mineera turned her head abruptly and stared into the darkness. She had caught the spiritual scent of Banion as he watched her and Tani from the darkness. Staring at his concealed location, Mineera considered the darkness with caution. She rose from the ground, stepping away from the fallen Tani. Holding her hands out in a gesture of submission, she proceeded cautiously. After formulating her thoughts, she began to speak, wanting to take the initiative.

"Welcome back," she said, startling both Jared and Banion where they stood, secluded in the darkness. Her gaze was strong and her speech confident. As the psychic stared into the darkness, Tani eyed her uneasily, not sure if Mineera had been gripped by another set of delusions, wondering to what or to whom she was speaking.

"Stay where you are." A commanding voice resounded from the darkness. Tani was set at ease. He could tell Banion was somewhere nearby, calling out to them. "Get over here, bookworm."

The young scholar obeyed the command and moved toward Banion's voice, though his exact location was still concealed by darkness. Emerging from the shadows with weapons at the ready, Banion and Jared advanced. Tani moved toward them slowly, wincing with pain as he moved. Coming to stand beside them, the battered scholar collapsed to one knee. Still weak and winded from the ordeal, he almost fainted.

The sights of the automatic weapons remained focused on Mineera as Banion sought to discover the truth about Tani's wounds. "Tani, what happened?"

"Guillotine tried to kill me," Tani said, still distraught about the incident which remained so fresh in his mind.

"Guillotine? Here?" Jared shot back in a confused tone.

"He ambushed me with a knife." Trying to avoid a fresh wash of tears, the frightened Tani looked away from his companions. Banion and Jared could see the raw emotion on the scholar's face and looked at him only briefly, not wanting Tani to feel uncomfortable; it was readily apparent that the tribal scholar was rattled.

Banion eyed the tribal's horrid wounds with concern. "How bad did he stick you?"

"Bad enough…" Tani said, his voice growing distant. The tribal was subsumed by a fresh batch of memories. The vicious thrust of the knife cutting his flesh was a chilling, vivid sensation which threatened to overturn his fragile sanity.

"My arms are cut up pretty bad. He stabbed my throat but missed anything vital. All in all, I made it out alive," Tani whispered.

"What about Mineera?" Banion asked aggressively.

"I just tried to…" she interjected, but Banion interrupted her with a fresh wash of anger.

"I am not talking to you," he said gruffly, madness flashing in his eyes.

Mineera fell silent. She could clearly see that she was still considered an outsider by the trio.

"She came to my aid, and just in time," Tani said, nervously watching both Jared and Banion's attitude toward Mineera.

"How convenient," Jared shot back, flushed with anger, staring at Mineera with mistrust.

"I am still not sure what to do with you…" Banion said, his voice ominous, while his eyes remained fixed upon her.

"Let's just get out of here. Leave her," Jared said in a frosty whisper.

"We had a deal, didn't we?" Banion addressed his words to the psychic, ignoring Jared's remark. "We had an arrangement that if you didn't get yourself under control, I would leave you out here, alone. Do you remember that?"

Mineera nodded in agreement. She had struck a deal with Banion before madness had taken her. Falling silent, she felt sorrow fill her. She had betrayed Nova 7, and it would take deep and generous compassion to bridge the void she had formed between herself and the others, especially Banion and Jared.

The conversation seemed to be leading inevitably to Nova 7 casting the psychic out. The thought of ousting Mineera, a strange prophet, but a member of Nova 7 nonetheless, alarmed the young scholar. It was a possibility that suddenly put Tani on edge.

"Wait a second," he objected. "She saved my life."

"She tried to kill Jared," Banion shot back.

"So we just leave her here?" The tribal was astounded by the lack of logic.

"In case you missed the show, bookworm, she attacked Jared. You remember that?"

"I'm with Banion. I say we leave her here." Jared was irritated, almost on the verge of yelling. Frustrated by the events, he was gripped by an adolescent fever of dissention. "How can we trust her after what she did?"

Tani was silent, not knowing what to think. On one hand, he didn't fully trust Mineera; on the other hand, he would have been dead without her. It was an indecisive place to be: happy to be alive, but afraid for his best friend. The group was silent as each one of them considered the recent events. Finally, breaking the tense silence, Tani spoke with an air of calm about him.

"I don't know how all of this is going to end. We have been out here, in the wastelands and ruins, for a long time. This bloody adventure is more dangerous than I was ever able to comprehend. In the blink of an eye, I almost died in this horrid place. I know more than ever how fragile we all are." Tani shook his head as if lost in thought and his gaze was distant. "See my wounds?" As he held his arms forward, the blood was still evident, clotted against his flesh in a maroon mass.

"I bear the signs of conflict upon my flesh. We all bear the signs of conflict upon us. Some of us marked in the flesh..." Tani held his bloodstained hand in the air for all to see once again. "Some of us bear the mark on our souls, stains upon our spirits." Tani eyed Mineera. Banion and Jared felt the wise words of the scholar sting at them as well. Each felt suddenly vulnerable as they stared at the bloody tribal youth, feeling ashamed for condemning one another. "We are all casualties of this quest." The words were profound, causing each of them to reflect upon their own demons.

"You remember how she used to be before Dune Station?"

The other companions of Nova 7 eyed Tani with uncertainty as he spoke. "Mineera had a glow about her, almost a sheen in her eyes. Anyone else remember that?"

"Whose side are you on?" Jared blurted out. He felt betrayed not only by Mineera, but by Tani as well. His best friend, essentially his brother, was taking sides with the person Jared viewed as the enemy.

"I am not on anyone's side," the tribal scholar replied, realizing that he was now caught in the position of protecting Mineera, his savior, from his best friend Jared.

"I can't believe you would protect someone like *her*." The warrior volleyed another scorn-filled comment back at his friend.

"You don't understand," he whined. "Guillotine would have killed me!"

"And I am supposed to believe that she came back just in time? How convenient!" Jared shouted, stepping forward, his cheeks flashing with rage.

"Calm down!" Tani was still weak, but remained intent on mustering a defense for Mineera nonetheless.

"Maybe I don't want to, Tani. You're supposed to be my friend. Do the right thing and take my side!"

"I am your friend. Give her a second chance. That is all I want you to do. Give her a second chance."

"Second chance? To do what? Try to kill me again?" Jared shouted back, anger etched in his face.

Banion stood silently, as did Mineera, watching Nova 7 tear itself apart. Tensions were running high and the debate was about to spill over and become even more heated.

"What was I supposed to do, Jared? Let Guillotine kill me so you can feel better? Look at me!" Tani's voice was hostile as he pointed at the wounds on his body. "I almost died!"

"So did I!" Jared shot back, glaring at Mineera. Frustration was turning to blind rage. The months and months of anxiety were erupting into a display of wild fury. They were all sick and tired of the stress and chaos, of escaping enemy agents, battling the violent forces of nature, and trying to keep from falling apart. The cost of the crusade had taken its toll on all of them. "I am so sick of all of this! You need to back me up. Can you do that, Tani?"

"I am always on your side, you should know that. But I am

also on Mineera's side. Let's just drop it…" Tani said, avoiding his best friend's eyes.

Jared, still flushed with anger, refused to give in. Shaking his head in disgust, the tribal warrior insisted on getting in the last word on the matter. "You know, for a *genius,* you are really stupid. Just stay away from me." Turning his back, Jared walked away from his friend, moving into the shadows of the abandoned tunnel.

The words stung painfully. Tani stepped back, looking at his friend in alarm. Jared's rage was uncontrolled, almost feral. Not knowing what to think, Tani shook his head back and forth, feeling betrayed. He turned away from Jared and refused to speak. The tribal warrior had crossed the line.

Mineera's homecoming was mired with ill will and despair. They were all demoralized, driven to the limit of human endurance both mentally and physically. Nova 7 was coming apart fast. The events that had transpired had turned the companions against one another. Banion looked at the ruins of his team and shook his head in dismay. The emotions were raw, and uncertainty was now the focus of Nova 7. Wanting nothing to do with the strife, the mercenary retreated into his own thoughts. The burden of leadership was all too real. If Banion didn't try to resolve the issues tearing them apart, the mission was in real peril. With a grumble, Banion shook his head again, close to despairing. Indecision and anger, he thought, now controlled the fate of the world by putting their crusade in severe jeopardy.

Chapter 6
The Race Begins

"You all don't get it, do you?" Banion called out, his tone harsh. The team was sullen and silent. The events of the day were still weighing heavily upon their minds. None of them had spoken in over an hour. Instead, they had all slipped away, hiding in their own thoughts. The tension was thick and oppressive as an impending winter storm. Insolent looks mixed with submissive glances which mingled with uncomfortable, aggressive gestures. Indecision and a flurry of emotions, rather than reason, were guiding the group. Instead of acting like a team, they were behaving like a moody collection of hormonal teenagers.

Tani was sullen, still hurt by Jared's angry remarks and upset about the violent ambush that had almost led to his demise at the hands of Guillotine. With his thoughts hovering between self preservation and loyalty to Jared, Tani's emotions were a wild see-saw, shifting from one mode of thought to the other. The scholar was suspended between his best friend and the woman who had saved his life.

Mineera was silent, still gripped with guilt. Her wild behavior had not only compromised Jared's life, but had put their crusade in serious jeopardy. If Mineera had not regained control when she did, Jared and possibly the rest of the team would have been killed, and the only chance of surviving the onslaught of the Reaper Kai empire would have been destroyed. Knowing full well that it would take time for the wound to heal and for the team to accept her fully once more, she resigned herself to keeping a low profile and waiting for chances to prove herself once more to the rest of her teammates. Following her plan of action, Mineera had been silent, wanting to lie low for a while until the emotional rift separating each of them from the others seemed to lessen.

Jared was hot-headed and didn't give a damn about anything at the moment, still feeling that his best friend had taken Mineera's side against him, and feeling betrayed. The once carefree tribal had been replaced by an ominous persona. There was a frantic, feral look in Jared's eyes. Where once he would have openly forgiven Mineera for her actions, his current temper would not allow such absolution. Jared was angry and wanted someone else to feel his rage.

Still lost in their own worlds, the members of Nova 7 were ignoring Banion. Irritated by their irreverence, he stopped and confronted all of them again. "Are any of you going to listen to me?"

"I'm not in the mood," Jared countered sarcastically, trying to push past the leader of Nova 7. Grabbing him by his shirt, Banion shoved the arrogant tribal back a step, his eyes brimming with madness. Jared gaped back at him, shocked by his aggression, but became docile in response.

"Not in the mood?" Banion snorted in disgust. "You truly don't get it! None of you!" Almost shouting, he looked at all of them as if to frighten them into listening to him. His demeanor was unnerving and achieved its goal; the members of Nova 7 emerged from their sulks and focused on him.

"Get what? What the hell are you talking about?"

"Guillotine. That's what I'm talking about. If he knows we are here, the enemy knows we are here. It won't take long for the Reaper Kai to put two and two together."

"We have been tracked ever since we left Rasheed. What makes this any different?" Jared shot back with a scowl on his face.

"In the past, enemy agents never made contact with us. There was no credible evidence of our location. With Guillotine escaping and making contact with Tani, the enemy will be on our heels quickly," the leader of Nova 7 responded. "We have a lead on a nuclear weapon. The enemy will undoubtedly send more agents and assassins to track us. We are now in a race with the enemy. If we cannot find the Runner in time, they will."

The gravity of their situation was beginning to register. It had suddenly sunk in. The elusive arms dealer, the Runner, the only being in the Darken Realm with a credible lead on a nuclear weapon, was the focus of their quest. Nova 7 would have to find the Runner quickly, or else the enemy agents would precede them. The prospect of the Reaper Kai recovering a nuclear weapon of their own was now a realistic threat.

Gripping his head with his blood-stained hands, the tribal scholar let out a sigh. "I never thought of that. I never imagined the enemy reaching our goal first. I simply thought they would try to stop us, not complete *our* mission."

Jared was silent, as was Mineera. The enemy could easily overtake them in this frantic race for nuclear power, thus ensuring the demise of the Darken Realm.

"We need to move. We no longer have the luxury of time on our side." Banion grabbed his gear, motioning the team to follow.

The other members of Nova 7 remained stationary, not wanting to look at each other or speak. A feeling of despair washed over them. The stress and strain of the past events had taken its toll on them. The thought of finishing their quest seemed like a distant dream, an intangible goal in an insane world on the brink of ruin.

"It's over, Banion." Jared shook his head in despair. "Let's just go home. I don't want to live like this anymore. Let's just go home."

The emotional tone among the team had turned to one of defeat. Jared's sentiment was shared by everyone; everyone except Banion. Feeling the blood rise in his cheeks, the fire exploded inside Banion. There was no way in hell the legendary gunfighter would simply surrender. Wanting to take the tribal apart with his bare hands, Banion shook in rage for a brief moment, trying to regain control.

"Get up and grab your gear," Banion ordered again, ignoring the tribal warrior's declaration of defiance.

Still they did not respond, each lost in their own thoughts, bitterness ruling them.

With a staunch look of determination, Banion gripped one fisted hand in the other. The team had fallen apart and if Banion wasn't careful, the nuclear crusade would end at this very spot, secluded under the ruins of a battered city.

In the past, Banion had been a loner, performing all of his missions solo. Though no longer alone, the burden of leadership was now his solitary duty. The team had hit an emotional low, and it was up to Banion to pull his team out of the mire. Unaccustomed to inspiring others, he quickly abandoned the notion. What the team needed was strong leadership, something in which Banion had almost no experience. With a stubborn rage flooding his senses, he saw nothing but the weakness of his companions. Falling back on his old ways, he decided he didn't give a damn about any of them. If they didn't tow the line and join him, he would finish the quest alone. Shaking his head resolutely, his fierce dark eyes washed over each of them, viewing them with evident disappointment.

"I'm leaving," Banion said, abandoning the thought of inspiring his companions once and for all. "Come with me or stay here, I don't care anymore. You can follow me and be heroes, or go home and live out the rest of your years as cowards."

With that, the leader of Nova 7 disappeared down a nearby tunnel. Though they were not intended to inspire anyone, his actions polarized the companions. His harsh, abrasive words had struck a chord with the rest of the mercenary team. Feeling somewhat foolish, each of them dug deep to dredge up another burst of emotional strength in order to continue the journey. With some hesitation, the rest of the team grabbed their gear and slowly plodded on after him.

Revitalized by a faint burst of strength, they pushed onward into the dark tunnels beneath the city which had been ravaged by nuclear fire in ancient times.

The team had only advanced a few steps down the dark tunnel when a low rumble rocked the ruins. Bracing themselves during the tremor, each member of the mercenary team steadied themselves as the tortured earth groaned. The walls of the tunnel began to crackle under the force of the miniature earthquake. Debris fell from the roof of the passage above, causing the companions to cover their heads with their hands. The tremor was minor and abated quickly, but reminded the companions once more how dangerous the ruins truly were. Sighing with displeasure, Jared scowled in the darkness of the tunnel as he voiced his grim opinion about the situation.

Banion was on the verge of madness. The bad attitude

enveloping his team was driving him toward anger at a constantly escalating rate. Shaking his head back and forth, he secretly wished that he could slip away and never return. He suddenly wished that his companions had given up and left him. If things didn't get better quickly, the enemy wouldn't need to end their crusade; they would do it themselves, torn apart by dissention.

The yellow flashlight beam barely illuminated the subterranean road ahead. Pressing on quickly, Banion managed to put some distance between himself and the rest of the members of Nova 7. It made his volatile mind seem a little clearer, even if it was only for a few moments. Coming to a junction, the leader of Nova 7 stopped and surveyed the choices ahead.

"Now what?" Tani asked sullenly, avoiding eye contact with Mineera and Jared.

"Not sure. The map is useless in this section of tunnels," Banion responded as he trained the light onto the poorly scribed map.

Mineera focused and let her powers stretch outward, probing the area for any signs of life. A flash filled her mind; it was brief but was substantial enough to get a visual impression of someone who had passed through the tunnels recently. In her brief vision, the traveler had taken the passage veering to the right. It was shaky at best but useful nonetheless. It was the only credible clue they had in determining how to escape the twisting passages. "I got an impression of the last person who passed this way. It's weak, but I feel he took the right hand tunnel." She spoke in a cryptic tone, her eyes still closed.

Jared snorted in her direction. "More of *your* friends that way? Maybe another ambush?"

Anger immediately flared again. The ill will of the team had reared its ugly head once more.

"Just leave her alone," Tani shot back, still angry about the last encounter between himself and the tribal warrior.

"I wasn't talking to you."

"Of course not, I'm too stupid, remember?" the scholar retorted.

"Quiet, both of you!" Mineera shouted, feeling disgruntled enough to yell. She didn't yell often but when she did, people took notice. For a brief second, the two tribals retreated, letting the matter drop. Shaking and with tears in her eyes, she came between them, hoping the bickering would stop. "You are friends! Stop this nonsense."

"Just stay away from me, both of you." Jared pointed at both Mineera and Tani.

"You done yet?" Banion grumbled.

"I'm not sure," Jared snarled.

Banion was enraged by his bravado. His eyes igniting dangerously, he took several quick strides toward the tribal warrior. The move was so aggressive that Jared took a step back in fear. Coming face-to-face with Jared, Banion stared him down. The tribal warrior broke eye contact first, shifting his gaze. "You are done, all of you are. Keep your damn mouths shut and follow me out of here. Anyone says anything, and I mean *anything*, I'm leaving you down here." His tone was menacing enough to stop the bickering, for the time being. The team knew his patience was gone, and none of them were crazy enough to test his temper.

"We have no clear option, so I am going with Mineera's suggestion," Banion said, still trembling in rage. "Let's go." With that, he took off down the right passage.

As Nova 7 slowly followed Banion through the maze of tunnels, evil was already plotting against them.

Guillotine was determined to contact his sinister employers. The mutant's goal was to notify the Reaper Kai command that he had made contact with Nova 7. News of Nova 7 in the Concrete Barrens would cause the enemy to immediately dispatch teams to hunt them down. With the location of the mercenary team no longer a secret, the Reaper Kai themselves would quickly set out on the trail of the nuclear weapon. The great race had begun, a race that would decide the fate of the entire Darken Realm. Only one victor could be crowned in such a race; either Nova 7 or the dread Reaper Kai would ultimately prevail.

Chapter 7
The Concrete Barrens

The steep slope angled upwards. Ascending the rough path, Nova 7 moved out of the secluded passages of the Iron Gate ruins. The dim nightmare of the underground was now behind them, a portion of the journey that all of them simply wanted to forget. Mineera's betrayal was still an open wound, tearing the team apart with pain and indecision. Though not openly feuding, the group was still filled with anger and dissention.

Emerging from the forgotten tunnels and passages, the companions now stared in awe at the once-mighty city.

The burning sands of the vast wastelands had flooded the ancient city. Dunes littered and lined the ancient byways and roads of the Concrete Barrens. The tan sand set against the gray concrete was an unusual mixture of color; the hue of the crumbling gray concrete created an odd, striking composition with the natural orange tint of the sand washing against the collapsed pillars. The companions felt as if they had stepped off the earth they all knew and found themselves on some alien world.

Rising from the orange sand were concrete skeletons, stripped bare by the passage of time and the nuclear fire in ages past. The sides of each building had been scoured. All of the glass and siding on the north and eastern portions of the buildings had disintegrated, leaving a melted conglomerate of glass, tile, and other materials. Like bones splayed from carrion, the steel rebar supports were twisted back, also melted by the heat of nuclear fire. Office furniture, ruined vending machines, household wares, gutted cars, old toilets, and other relics were scattered around the perimeter of each building, partially obscured by the shifting sands.

There was a slight chill in the air of the ruined city. Moist fog from the nearby ocean to the west lingered amongst the buildings. Like ghostly tendrils, the eerie mist changed form, drifting between the skeletal remains of the buildings. The tall monoliths of concrete rose into the fog, so high that their tops could not be seen in the thick haze.

Unbeknownst to the team, hundreds of such ravaged sky scrapers, miles upon miles of them, stood silently in the dreary fog like broken coffins littered amidst a mist-filled graveyard. Near the center of the ruins was a barren patch of earth, still blackened and radioactive, a place that not even the fog would dare touch, a wound so horrid that it reeked of death and always would. With hundreds of thousands burned to ash in a blinding flash of hate, the blast crater was a monument to technology and morality gone awry.

The team stood in the white haze, looking around them, feeling small and insignificant. It was a lonely sensation, to stand in the abandoned streets that had teemed with life eons ago. Whether it was the frosty breath of the fog or the loneliness of the ruins chilling their souls, the team felt isolated, and an icy fear tickled their fragile sensibilities.

As if waiting for the city to greet them, the members of Nova 7 stood silently, viewing their surroundings in awe.

The city groaned and the earth heaved. A slight tremor rocked the ground. Just north of their position, a spike of fire erupted from the earth. The orange flare rose into the air, tumbling as molten lava burst forth. Like a splash of sluggish clay, the superheated magma rained down, withering a nearby building further by partially melting a brick wall. Enraptured by the display of raw power, the team members required a brief second to recover their senses and focus on the quest at hand.

"Let's get moving," Banion ordered gruffly. "We need to locate the Runner."

"This Runner is the legendary arms dealer? Living here? You have got to be kidding me," Tani concluded, staring at Banion. The deranged gunfighter looked back with a hostile glare.

"This is the place," Banion responded, but continued to stand still, ignoring his own command. He seemed confused about their next course of action.

"Where to now?" Jared quizzed Banion, still refusing to look at Tani or Mineera. To the tribal warrior, they simply didn't exist.

"Not sure," he shot back, increasingly irritated by all the questions. "Can you just be quiet for second so I can think?"

Banion surveyed the ruins. All directions looked pretty much the same; each path was an avenue littered with the decayed ruins of an ancient empire. Taking out the map he had received from the Consortium of Arms, Banion took a stab at gleaning additional information from it, to no avail; the map did not contain any information beyond that provided about the underground passages leading into the Concrete Barrens. Shrugging, he gave up on the map and took a moment to orient himself. "From the ridgeline in the mountains above the city, it looked like the eastern portion of the city had been destroyed by a nuclear blast. I say we push west. Any local population would have to live outside the radioactive crater." He turned around to look at the rest of the team, gauging their reaction.

Much to his dismay, they were still bogged down by a shoddy attitude; none of them were too keen on acknowledging their leader's decision. Tani shot Jared an apologetic look, still feeling guilty about standing up for Mineera. Jared refused to look back, still scorning his friend. Mineera was lagging near the back of group, trying to remain inconspicuous, still dealing with the heavy burden of her own guilt.

The group was still in sad shape. Banion shook his head in disgust, seeing the indecision and conflict gripping his team. While Banion was bold and rash, he still needed to acquire some experience in inspiring his teammates and dealing with the conflicts between them. For now, he would ignore their plight, preferring to heal the wounds at a later time. "Anyone else alive here? I said, I think we should go west," he repeated.

Silence was the only response. Sighing in frustration, Banion turned westward and began walking. The rest of the group followed silently.

Nova 7 pushed deeper into the ruined city. As they moved further into the battered metropolis, the fog grew denser, chilling the companions to the bone. In the dim light and shifting mist, images seemed to take shape, then vanish.

"I keep seeing things… in the ruins… lumbering in the fog…" Tani stuttered out. On more than one occasion, he had thought he caught sight of shapes shambling at the edge of his vision; in the blink of an eye, the images would vanish, leaving the tribal scholar blinking several times, trying to focus on the fading glimpses of movement.

Jared spun around, scanning the rubble behind them. He, too, thought something was moving behind them in the white haze. "I agree, there *is* something out there." As Jared's voice trailed off to a whisper, a crash rose from the fog, distant but clear nonetheless. It sounded as if a rusted metal garbage can had been knocked to the ground. More sounds, very slight and at the outskirts of his hearing, caught his attention and set him further on edge. His hand instinctively dropped to the hilt of the mighty Scar Blade.

"Be on your guard, all of you." Banion held the assault rifle to his chest, readying himself for conflict. His gaze narrowed and he clicked off the safety on the weapon.

"We are surrounded," Mineera whispered in a rasp, her voice still echoing through the white fog. The psychic could feel life all around them. On every side, she caught wisps of emotion emanating from the mist. The awareness she sensed was diffuse and erratic but seemed intelligent in nature.

"Grab your guns," Banion ordered, feeling as if an ambush was only seconds away.

Jared drew his submachine gun, as did Tani. Holding the weapons ready, the tribals prepared to face whatever was concealing itself within the ruins. For the moment, the threat of danger had brought the team back together. For now, there was no thought of mistrust, only the spike of fear in their hearts, giving them focus, uniting them with the goal of survival.

"Keep moving," Banion commanded, leading his crew deeper into the ruins and further into danger.

The team passed only a few city blocks before they came across a large open area, probably a city park in times long past. As they moved into the area, the mist and fog seemed to heave upwards and a light breeze whispered against their skin. The fog swirled and was pushed upwards into the sky, liberated from the skeletal remains of the shattered buildings by a rising thermal updraft. The dense fog abated near the center of the open area and Nova 7 gasped in fright.

To say they were surrounded was an understatement. As the fog lifted, the members of Nova 7 found themselves staring at a shambling mass of thousands of misshapen humans. In the open area built upon the shifting sands and dunes flooding the city was a shantytown. Garbage and materials scavenged from the ancients had been thrown together into a collection of crude buildings. The shacks were densely packed together in the improvised settlement.

Once the fog had dissipated, the inhabitants of the shantytown caught sight of Nova 7.

They streamed from the open doors, hunching forward in a mass of lost hope like an army of zombies on the move. Many of them sniffed the air, catching the scent of the newcomers. Their white eyes bulging, the sickly human horde moved to confront the hapless explorers.

The citizens of the shantytown were a mess. Poorly dressed in rags, the wretched rabble crawled forward. With jerky movements they advanced, their pale white skin speckled with sores. Blank expressions covered their faces as they staggered forward slowly, like a mass of the dead on the march. Taking a step back, Banion held his weapon ready. The tribals did the same as the wall of sickly humans jerked forward.

"Slumlanders…" Banion whispered as he stared at the sickly humans moving forward.

"What?" Tani mouthed.

"A Slumlander is a person that lives in a contaminated region. They suffer from sickness and mutation."

"We need to get out of here," Jared said aggressively, watching the shambling human horde advance.

Spinning around, Nova 7 surveyed the tall buildings around the perimeter of the shanty town. More wretched forms were clambering on each floor of the ruined structures. Hundreds more stared down from the tops of the buildings. They were everywhere, an ominous mob of sickly humans. Panic set in. They were surrounded, outnumbered, and in an unknown environment. With each passing second, dozens of additional sickly humans shambled into view, a living wall that would take a miracle to break through. With tension rising and thousands of Slumlanders lurching forward, Nova 7 prepared for the worst.

Chapter 8
A Deal is Struck

The assault rifle's trigger twitched nervously under the weight of Banion's index finger. Banion stared in horror as the horde of shambling humans, misshapen from mutation and disease, lurched forward with blank, vacant looks. The situation was grim, and the leader of Nova 7 was out of options. The team was fully surrounded and about to be overrun.

Sizing up the situation, Mineera knew that fighting their foes would mean a certain death, as Nova 7 was outnumbered several thousand times. The rest of her teammates were stunned and could not seize on a clear course of action. Knowing that the encounter could end in their demise, Mineera took the initiative and formulated a plan of survival. Coming to a decision which might prove rash, but which seemed to be a viable means of escaping conflict, she ordered the rest of her teammates, "Put your weapons away."

The other three members of Nova 7 were far from receptive to her plan of action.

"No way," Banion shot back vehemently.

Only twenty yards away, the living wall of shambling humans continued its advance.

"We cannot win this fight. Look around you. We are outnumbered." Mineera pointed up at the buildings towering above them. The rest of the team spun around and caught sight of several hundred more of the sickly humans staring down with blank, hungry looks. "I don't sense that they are hostile to us. They only seem suspicious. Trust me," she coaxed in a persuasive tone. "Put your weapons away."

Banion looked at the mass still hunching forward. The sores upon their pale white skin drove a wedge of fear into his mind. Shaking his head with indecision, Banion turned to Mineera. "I hope you're right." The gunfighter was taking a big chance by lowering his weapon, but in the end he agreed with Mineera; there was no way that hostility could be advisable when outmatched four thousand to four.

"Shoulder your weapons," Banion ordered, and the tribals shouldered their submachine guns with hesitation, wide eyed with fear.

The mob paused in their tracks; rank upon rank of pale, sickly-white humans stared at them with sunken eyes wreathed with black sockets. The starving humans eyed them without blinking, and the moment stretched on, growing almost unbearably strained. And so they stood, waging a taut stalemate. The members of Nova 7 looked back intently at the locals, trying to size them up and discover their intent. The confrontation was tense and neither side made a move. Finally, there was movement at the back of the ranks of Slumlanders. Someone was pushing through the mass.

A strangely elegant man made his way through the crowd to confront Nova 7, stationing himself in front of the shabby army of shambling humans. Dressed in an exquisitely crafted tunic and luxurious cloth pants, he was unmarked by disease, and seemed to be the leader of the Slumlander tribe. A white goatee flowed from the edge of his chin almost down to his waist. He was totally bald save for a long, braided sprig of hair erupting from the top of his head. His dark eyes stared at Nova 7 with mistrust. Holding his arms behind his back, he moved forward to address the mercenary team.

"What brings foreigners into our territory?" The Slumlander leader spoke with a flare of anger rising in his voice. Mineera could sense his feelings and knew that this interaction required caution. Stepping forward and positioning herself in front of Banion, Mineera took the initiative and decided to confront the Slumlander leader herself. The action was bold and unnerving to the rest of Nova 7.

Confounded by her bold assertion, Banion stared at her in anger. The decision to confront the Slumlander leader was aggressive and brash. It showed that Mineera had regained her confidence – a little too much confidence, in Banion's opinion. But ultimately, he decided to step back and allow the seasoned diplomat to work her magic.

Coming before the finely dressed Slumlander, Mineera bowed in a gesture of respect toward the leader. Her gracious display of civility was well appreciated, and the Slumlander leader bowed back.

"I am Mineera, and these are my companions," she said calmly, maintaining unwavering direct eye contact with the leader.

"I am Prefect Dale, leader of the Concrete Barrens." He peered back with an unreadable look upon his face. "From where do you and your companions hail?"

Not flinching for a second, Mineera replied in a calm and confident tone, "We hail from the city of Rasheed, far to the east of here."

"Ah, Rasheed. I have heard tales about the mighty city far to the east. What brings you into our home?"

Mineera hesitated to tell the Prefect the intent of their journey, speaking only after she received Banion's nod of permission. "We are searching for an arms dealer known as the Runner."

As the words rolled off her tongue, the crowd surrounding them was infused with a fresh wash of anger. Prefect Dale scowled at her with a look of mistrust, and Mineera picked up on his feeling of anger. Knowing that one more ill-chosen move could mean their death, she moved to diffuse the situation. Discerning that the anger was associated with the topic of the Runner, she sought to create a ruse to bring the Slumlanders back to their side.

"Why are you seeking the Runner?" the Prefect growled with hostility.

Planning the ruse carefully, Mineera began to spin a vague tale about the Runner. "We have been hired to contact and *deal* with the Runner."

"*Deal* with him?" The Prefect suddenly became interested. "What do you mean, *deal* with him?"

"To protect the interests and safety of our employers, we have been commissioned to acquire certain possessions from this 'Runner'. Our orders are to intercept the goods by *any means necessary.*" She was coy and calm, subtly stressing that the Runner was not an ally.

Prefect Dale considered her with a stern gaze; Mineera never wavered and never averted eye contact. The Prefect was buying the story, and his level of anxiety was dropping away quickly. Eyeing both Banion and the tribals, he seemed elated. The quality of weapons the team exhibited was a good indication that Nova 7 was accustomed to conflict and battle.

Smiling, the Prefect eyed their guns with hungry eyes, softly repeating the words Mineera had just spoken: "*By any means necessary...*"

As he emerged from his greedy trance, Mineera could feel a veneer of deception wash over the Prefect. He was becoming emotionally evasive, and this put Mineera on edge. Smiling as if attempting to conceal his excitement, the Prefect of the Slumlanders asked a critical question. "Do you know where the Runner is?"

Mineera felt as if she was walking into a trap, but had no other choice but to try and ascertain the location of the Runner. "No. Our orders did not specify his location other than this city."

The Prefect smiled, and Mineera could feel a tinge of evil emanating from him. Something was not right. Her senses were clear and crisp, and Mineera knew that the Prefect was seeking to manipulate her and Nova 7. She considered him cautiously.

"I have a deal for you, then," Prefect Dale said, still smiling.

"What kind of deal?" Banion shot back, taking a step forward. The wizened mercenary could smell a trap as well, and didn't like it one bit.

"Our tribe is waging a costly war with the environment here. Sickness is killing my tribe slowly. My people are in need of clean water. Our purifier has broken down and we require a replacement part. We sent several expeditions to the ocean west of here, where the once-mighty city grew proud. In ages past, the earth broke, and much of the city sank into the ocean. A series of islands still remain, not yet claimed by the sea. Our plan was to scour the ruins of old for suitable replacement parts. From our last expedition, only one of the five we sent returned, badly injured, on the verge of death. Horrible wounds covered his body..." The Prefect spoke as if in a daze.

"What happened to your scout? Did he survive?" Mineera

quizzed, feeling a sense of dread surrounding the failed expedition.

"No…"   Shaking his head, the Slumlander leader remembered the scout's wounds. "It was as if his flesh had been seared or burned from his body. I can still see him…" The Prefect's tone was grim. "The skin hung loosely, as if the man was boiled alive by some horrible force. I have never seen such a thing." Shaking his head in dismay, Dale snapped out of the grisly daydream. "That was the last expedition we sent. The scout, in a state of shock upon his return, spoke of horrid creatures living upon the islands. Before he died, the scout stated that the expedition had located the replacement parts, the purifier parts we needed. After their discovery, that's when they were attacked."

A moment of silence washed over the group. Mineera held her tongue while the Slumlander leader contemplated the grim circumstances once more. When Mineera felt his spirit to be more at ease, she pressed on with her own thoughts. "I am saddened by your loss. Do you know the location of the parts?"

"No, the scout died from his wounds before divulging the location. All I know is that somewhere out on those islands, in the ruins, are the purifier parts we seek." Stroking his long white goatee, the Slumlander narrowed his eyes as he spoke to Mineera. His tone grew suddenly aggressive. "The purifier parts on the islands are a long shot. Fortunately, we have a definite location for another set of replacement parts. Unfortunately, the only other replacement parts are held by a greedy neighboring tribe that lives here in the Concrete Barrens." The Prefect was concocting some sort of twisted deal; Mineera and Banion were both on edge. "If we could only obtain these parts, our strife could end, and we could live out full lives, avoiding contaminated drinking water."

"So what is your deal?" Mineera asked firmly, knowing full well where the conversation was leading.

"We have been fighting an open war with the neighboring tribe for a long time to secure the parts we require. The bloodshed has been costly for both tribes. All we require is for you to sneak into the enemy camp and liberate the purifier parts we require. In exchange, we will give you the location of the Runner in this city."

A silence washed over Nova 7 and the Slumlander tribe. Banion and Mineera turned around to consult with Tani and Jared. They spoke in hushed tones, not wanting the Prefect to hear their words.

"I don't trust him. He's hiding something and seeking to manipulate us," Mineera said.

"I don't trust him, either," Banion concluded.

"Do we have a choice?" Tani asked. "We can either locate the parts in the ocean ruins or steal them from the neighboring tribe. It sounds like all of their expeditions failed."

"We don't just have two choices, we have three. We either deal with this guy, search the ocean ruins, or we wander around the city looking for the Runner ourselves," Jared interjected. "And with Guillotine stalking around, can we afford to waste a moment's time? If we don't deal with this guy, it will take us a while to locate this Runner. If we take too long, this whole area will be swarming with Reaper Kai and more enemy agents."

Banion nodded in agreement to Jared's rationale, concealing a brief smile. Nova 7 was acting like a team once more. For the moment, the group was at peace, focusing on the issue at hand as opposed to their own selfish conflicts.

"Can you sense if he even knows where the Runner is?" Banion asked Mineera.

"Yes, he definitely knows where the Runner is, but there is a high level of animosity around the topic," Mineera whispered, then sighed and shook her head. "If I could make a decision based on my feelings, I would conclude that the Slumlander leader hates the Runner. I smell a trap."

"I don't think we have a choice," Tani responded. "We can't risk another ambush by Guillotine. We must find the Runner as quickly as possible. I don't trust him at all, but we don't have a choice. Time is our enemy. I say we take his deal."

"Tani is right. This city is too dangerous to simply wander around in. Each day we waste is a day given to the Reaper Kai. If we don't find the nuke, the enemy will. We have to make a deal and hope we can avoid whatever this Slumlander throws at us," the leader of Nova 7 concluded. "Tani, you think you can figure out how to remove the parts from the water purifier?"

"Yeah, it shouldn't be a problem," the scholar said confidently. "I just need to know what I am looking for and my tool kit."

"Whoever we steal these parts from is going to be angry.

We are going to destroy their water purifier to fix the Slumlanders' purifier. I hope we don't get killed in this whole mess," Jared concluded, and the others agreed.

Banion sighed. "The best course of action is the most direct route to the Runner. I think we should steal the parts from the neighboring tribe. The enemy's agents will be close on our heels after the encounter with Guillotine. We must be careful and do this quickly." Nodding, Banion made his decision and turned to Mineera. "Make the deal."

Mineera approached the Prefect. "You have a deal. We will get you the parts in return for the location of the Runner. All we require are the damaged parts, so we know what we need to find."

Smiling broadly, the greedy Prefect was clearly elated. "I will draft you a map to the enemy stronghold. We will also provide you with the broken parts so you know what to look for."

The deal had been struck. Nova 7 was now entangled in a sinister plot with the Slumlander tribe. Their goal was constantly getting more complicated, the deal they had just made was a desperate one, and Nova 7 was sinking deeper into a sinister plot.

As the agreement was finalized, they surveyed the mass of shambling humans all around them. The haunting image revealed to them in the rolling mist would forever sting the minds of the team members: a horde of pale white humans rotting to death in a forbidden city at the edge of the world.

Chapter 9
Delving into Darkness

"Why are you still trying to talk to me?" Jared snarled, staring at Tani with contempt. "I told you to leave me alone."

The tribal scholar's expression was sheepish. Frustration had taken over once more. Slouching like a defeated soldier, he let out a sigh and rolled his eyes at Jared. The trio had been bickering again. The ill will and bad attitude were still a threat to the team, and the situation was beginning to get out of control. Their primary task, locating the purifier parts, seemed to pale in comparison to everyone's constant attempts to get in the last word, or one last crude comment at another's expense.

At last, the rest of the team noticed Banion's hostile glare and decided to put aside their differences for the time being.

The companions stood at the edge of an open pit leading into the bowels of the earth. Banion stared down silently, as did the rest of Nova 7. His expression perturbed, Banion turned to the map given to him by the leader of the Slumlander tribe. Shifting his eyes back and forth, he grimaced in disgust as the reality of the situation became all too real.

"You can't be serious," Jared said with a hint of disgust in his voice. The tribal warrior was staring down into the darkness of the open pit.

Tall buildings surrounded a raw wound which had ripped into the ground. The earth in this portion of the city had collapsed, swallowing all of the buildings in the near vicinity. What remained was a gaping hole. Sand from the desert had flooded the streets of the city, and small grainy streams continually poured into the opening, drizzling into the darkness below. Watching the tan granules trickle into the foul pit and disappear into the darkness was somewhat unsettling to the team. It almost seemed as if the hole in the earth was a bottomless pit, trying to consume what was left of the world above.

If the gaping void wasn't enough to deter them, the odor rising from the pit was strong enough to drive almost anything away. Even as they held their noses, they could still feel the stench sting their nostrils. A wretched smell of death and noxious chemicals wafted from the opening.

"No way," Tani said, shaking his head back and forth. "This is crazy. I don't want to go down there, into that hole."

"We don't have a choice. If we don't get the parts they need, the Slumlander Prefect won't help us locate the Runner. No Runner, no nuke. No nuke, the Reaper Kai win," Banion responded, glaring at both Jared and Tani.

"I sense immense spiritual energy emanating from the pit. There is great evil down there. It's diffuse, almost like a whisper, but it's strong nonetheless." Mineera spoke in a haunting tone. Her blue eyes opened gradually, drowsy and opaque; she had just scanned the darkness with her formidable powers. "I don't think this is a good idea."

Banion was becoming hostile. His plan was beginning to crumble, and his obsessive need for retribution was beginning to mount. Not wanting to surrender, he pushed further to get his way. "You want the enemy to win?" he asked the rest of Nova 7, his tone confrontational.

"That's not it." Jared remained defiant. "The whole deal stinks. That Slumlander wants us to *steal* working water purifier parts from another tribe. We have to destroy another tribe's water purifier to save the Slumlanders. It doesn't make any sense. We have to ruin another tribe to save the Slumlanders. I thought we were above that."

"Hey, kid, sometimes you need to make hard choices and sacrifices. Especially in the situation we are in."

"Yeah, but we wouldn't be in this situation if you hadn't ordered Mineera to cut a deal with that maniac," Tani added with venom in his voice. The young scholar pushed his glasses up his nose, noticeably irritated. "We could have searched the ocean ruins."

"The ocean ruins? You heard the Prefect. Several expeditions died out there. I would rather take my chances and steal purifier parts from a known location than search around in dangerous territory, especially with Guillotine on our trail." Banion had little patience to argue his decision with the scholar.

"I felt extreme deception from the Slumlander leader. He will betray us." Enveloped by her blue robes, Mineera moved forward to confront Banion. The team was coming apart fast. The bickering was beginning to erupt once more, causing Banion to shake his head in frustration. What Banion needed was a way to motivate his team, get everybody on the same page. If the mistrust and bickering continued as they had over the past few days, it would put their mission in jeopardy.

"Let's go down there, get the parts, and put this all behind us. Tough times call for tough decisions. If I had another choice, we would be doing it," Banion said gruffly.

Shaking their heads, the rest of the team agreed uneasily. Without locating the Runner, their goal of finding a nuclear weapon would be unattainable. And with Guillotine still tracking them, time was of the essence. Sighing, Nova 7 prepared to descend into the foul-smelling pit.

Moist, humid air rose from the opening of the pit. The rising plume was a strange mixture of rotting flesh and medicinal herbs. The distant depths of the chasm were devoid of all light. The sunlight barely lit the chasm even a mere thirty or forty feet down. The only sound reaching their ears was the constant trickling of sand, sweeping off the dunes that had covered the abandoned city streets and raining into the hole. The scene was ominous and put them all on edge, filling them with an intense fear of the unknown. What hidden dangers lay within? What horrible fate could be in store for Nova 7?

Hiding their fears, they prepared to enter the foul abyss.

Banion knelt and tethered a rope, securing it to a fallen concrete support. With a fleeting look of accomplishment, he threw the rope down into the dark hole. There was a silent moment of anxiety as the rope descended into the void. In a few seconds, it hit bottom with a thud. The team was elated; the rope had not fallen too far – the bottom of the pit was closer than they had originally thought.

"Here we go," the seasoned gunfighter declared with a hint of displeasure.

Banion was the first to descend. Grasping the rope, he slowly pushed over the edge. Pressing his feet against the wall, the veteran mercenary slid downward into the darkness. After about seventy feet, he hit bottom. Groping around with his feet, he felt secure ground beneath him and let go of the rope. Lighting a flashlight, Banion secured the small light source under the barrel of his assault rifle. Panning the weapon around, he surveyed his surroundings.

Banion was standing on a mound of shifting sand. The dunes had been filling the ruins for centuries. All around him, the ground was covered with a thick layer of orange sand.

Sweeping the space with his flashlight, he caught sight of a large, ornate doorway just west of his position, assembled out of carefully stacked, light tan sandstone blocks. Chiseled into the finely engineered doorway was a series of sculptures. Still at a distance and unable to see the inscriptions, he moved forward to inspect the doorway. Banion surveyed the inscriptions with growing shock. A sinister collection of skeletons and skulls covered the perimeter of the passage. There were dozens of such pictures, depicting inhuman skeletal beings, with empty eye sockets carved into the stone. The sockets were deep enough to cast shadows, making the skulls appear to have blackened eyes. Banion gasped at the sight; it was so unexpected that it caught him off guard.

"Hey! You down there?" Jared shouted into the hole, still standing at the top edge of the pit.

There was a moment of silence. Banion was still mystified by the sinister doorway adorned with skulls, skulls that did not look human. The bone structure in the sculptures depicted the skeletal remains of a creature with an elongated snout containing a jagged row of sharp teeth. Flashes of memory filled Banion. In that instant, his thoughts drifted back to the encounter that took place many moons ago in Shape Home. That night, Banion had encountered strange creatures within the Consortium of Arms, the same arms dealers that had sent him all the way into the hostile Concrete Barrens. The creature with whom Banion had struck a deal was cloaked, its form obscured, but it had an elongated nose. A wash of anxiety filled him as he stared at the familiar shape of the skulls adorning the passageway. There was a strange resemblance between the gun runners and the beings on the sinister doorway.

"Banion?" Tani shouted into the dark hole.

"Yeah, come on down," the gunfighter shouted in response, the hair on the back of his neck beginning to stand on end. Crouching, he held his gun tightly, pointing the barrel toward the tunnel beyond the doorway. He was fearful of what might lie inside and was determined not to be ambushed by anything hailing from the tunnels beyond.

Jared was the first to reach the bottom. He steadied himself on the mound of sand. With a gesture similar to Banion's earlier one, Jared affixed his flashlight under the barrel of his submachine gun.

"Secure the other side of the area, kid," Banion ordered Jared, and the tribal warrior pushed into the darkness on the other side of the passage. The sandy floor shifted quickly under Jared's feet as he moved to secure the bottom of the room. Panning his gun downward, Jared felt fear rise at the bottom of his stomach. The sand, disturbed by his movement, was falling away, deeper into the earth. The eastern side of the room had been fractured by an earthquake. A ledge was just ahead; the dislodged sand was falling off into a black abyss. The gaping void was deep, very deep, so vast that the tribal's small flashlight beam could not reach the bottom.

"There's nothing to secure over here but a huge pit." Jared spoke softly, looking into the abyss with a shudder.

Descending rapidly, Tani and Mineera reached the bottom. In the dim light, the two newcomers looked around in confusion. That moment of uncertainty was when it happened.

A small tremor rocked the ruins. The rumble rose through the darkness as a minor earthquake rocked the tunnels. Crackling, the stone walls heaved. The ground rumbled and rocked. All of the companions braced themselves, feeling their balance shift quickly. As they tried to find solid ground, each of them lost their footing.

Jared, still near the edge of the deep pit, steadied himself. As the tremor continued, the sand slope the tribal warrior was perched upon began to shift quickly. The mound collapsed, sending the slope of sand into the deadly pit below. The tribal yelped as the earth under his feet gave way, sending him toward the edge of the pit with frightening speed.

The youth rolled down the slope and with an agile move, his hands latched onto the ledge as his feet fell into the void below. The tribal's body hung over the void as he grasped desperately at the ledge. The gun strapped to his chest spun around, winking rapidly in the dim light.

"Help!" Jared yelped, his submachine gun dangling from his neck, his legs kicking wildly in an attempt to gain a foothold. To no avail; the tribal could not touch anything with his feet. Panic rushed through him. Adrenaline flooded his senses as he fought to keep clinging to the narrow ledge.

The tribal's fall stunned and temporarily paralyzed the rest of Nova 7.

"Do something!" Tani shouted at Banion with a frantic whine.

As Banion looked at Jared dangling over the edge, a smile graced his lips. Twisted gears within the gunfighter's brain began to rotate and grind away. The strange circumstance could lead to Jared's death, but to the unstable Banion, the event was a lucky one. Within a split second, he had formulated a mad plan. Instead of seeing the event as dangerous, Banion saw it as a way to bring his fragile team back together, allowing his companions to act like a team instead of a bunch of whining, out-of-control brats. With a sinister smile, Banion declared, "I'm not doing a damn thing. You want to save him? You and Mineera do it."

"What?" Mineera cried in disbelief, frozen in terror.

"Help me!" Jared yelled again. His fingers stung as they gripped the ledge. Even though the youth was strong, his stamina was beginning to wane.

The horror of the situation set in. Jared was hanging on for dear life and Banion was doing nothing about it. Tani and Mineera looked at each other and rushed down the slope to save Jared.

They reached the edge in the nick of time. The tribal warrior's strength had faded. Both of them grasped an arm, holding him above the void. With blood pumping and fear driving them, Mineera and Tani pulled with all of their might. The tribal warrior was helpless, utterly at their mercy. A tense battle was waged as Mineera and Tani harnessed every iota of their strength, their muscles shaking with fatigue under the tribal's weight. Finally, their efforts were rewarded and Jared was pulled to safety.

Gasping, the trio lay in a huddled mass upon the sandy slope, piled on top of each other in a single heap. The rush of adrenaline was fading. Tani, Mineera, and Jared looked at each other in wide-eyed fear, still unable to comprehend what had just happened. Banion came down and crouched beside his teammates with a serious look on his face.

"What the hell are you trying to pull, Banion?" Tani lashed out at Banion.

Jared was rigid with fear, unable to believe that he was still alive and had not fallen to his death.

"I saved your lives," Banion said in a triumphant tone, a look of madness in his eyes.

"Saved our lives?" Mineera's voice was filled with quiet rage. "Your inability to act almost killed us!"

His dark brown eyes seemed to simmer in the darkness as he flashed them a crooked grin, considering them all with pity.

"If we do not work as a team, we are all dead." He looked at each of them in turn. "For the past few days, you have bickered and squabbled. Mistrust and anger have ruled you. Our team has fallen apart. We have lost the ability to trust each other. Without that trust, all is lost."

"You would have let me die?" Jared asked in a shocked tone.

Staring intently at Jared, Banion responded. "Yes, I would have let you die. If not today, it would have been soon, anyway. You are all on a reckless road. If you don't act like a team, if you don't trust each other, we'd all die anyway. The bickering and anger you have toward one another would have destroyed us. The enemy shows no mercy. The weakness we have turned against each other would have allowed our enemies a swift victory." He spoke methodically, and his logic was chilling. "So yes, I would have let you die. I would have let you die to save our team. In the end, none of us can do this alone. If we cannot find strength in each other, we will find weakness divided."

The statement was so profound, it hit them all powerfully. Banion's words rang true. It all made sense. If Nova 7 could not trust each other, it would be impossible to finish the mission.

Without absolute trust, the agents of the enemy could drive them apart with fear and kill them one by one.

"I want you all to look at each other," Banion said forcefully. "These are your companions. Today, the anger ends. From this point forward, you will trust in each other absolutely. Do you all hear me?" he asked, extending his open hand into their midst. All of the remaining team members were stunned, looking around the group and feeling the power of Banion's wisdom. One by one, they all agreed, and each of their hands joined Banion's own. With the burden of animosity lifting, the newly reunited Nova 7 felt their spirits leap. There was no reason to bicker amongst themselves; instead, a feeling of empowerment coursed through them and smiles illuminated their faces. The broken team had been remade.

Looking at the entwined hands linking them together, each member of Nova 7 felt strength return. No longer would dissention rule them. The past was over, and all misdeeds were forgiven.

Each of them marveled at Banion's heart and courage. His decision not to save Jared from the pit had actually pulled them together. It was an insane twist of fate, but the event was profound. Nova 7 was a team once more.

Standing before them, Banion grinned and asked, "Are you all with me?"

With the fire leaping in their hearts, their spirit had been reforged. A bolt of power filled them. No longer would hate and mistrust rule their destiny. Smiling back at him, the rest of Nova 7 nodded that they were indeed with him.

Another flash of madness lit Banion's face. He spoke once more, his tone mischievous. "Let's go steal those water purifier parts."

With hope filling them, the mercenary team prepared to enter the tunnels surrounded by pictographs of the dead, visions of skeletons carved in the very walls of the sinister passages beyond. Their resentment gone, Nova 7 stood united, and united they needed to be; they were about to uncover one of the most carefully hidden secrets in the entire Darken Realm.

Chapter 10
Plunge into Darkness

The red etched sandstone was covered with a host of inscriptions. Unknown words in a mysterious language were chiseled into the rock wall, overshadowed by the sinister pictorials delicately carved into the red stone. The skeletal depictions of some unknown mutant species were covering the walls and ceiling in regular intervals. In their extreme intricacy, the pictorials were a monument to an advanced society.

Nova 7 was silent. Each member of the mercenary team was shining their flashlight along the walls, surveying the pictures with fraught anxiety. Banion's beam, emanating from the flashlight attached to the barrel of his assault rifle, fell across the largest of the scrawlings, a wretched form covering the archway leading into the dark passage beyond.

The scrawling upon the entryway was a hunched and skeletal form. Each teammate held their breath as they drew near, eyeing the finely carved skeleton of a beastly mutant. The strange creature was squat and curved, almost as if it had been compressed in some way. Its spinal column was twisted, making one think the monster could push itself into any opening. The skull of the beast was especially foreboding. Small sockets were set upon a skull with an elongated snout filled with a toothy maw of jagged teeth. It looked as if a saw with dozens of serrated points had been inserted into the jaw of the creature chiseled upon the wall.

A silence hovered between the group members. Their internal dread mounting, they all knew that whatever lay beyond the archway would test their courage. Nova 7 was well beyond safety; instead, they had plunged themselves into one of the worst places in the entire Darken Realm.

"Jared, get over here," the leader of Nova 7 ordered. Jared responded obediently and moved right beside Banion. The seasoned mercenary was training his weapon into the darkness of the open archway. His flashlight beam traveled down the tunnel a mere thirty yards before the dim beam was eaten by the shadow of tunnel. "Cover this doorway with me," Banion instructed.

Jared obeyed and readied his submachine gun. Trying to avoid looking at the pictorials, he was unable to avert his curiosity. Finally, he gave in, darting quick glances around, and was stunned by what he saw. The sinister animal skulls carved into the open passage were unnerving. The tribal blinked several times, thinking that he was seeing some sort of illusion.

"Wonderful," he declared sarcastically.

"Bookworm, grab your gun and get over here." The leader of Nova 7 was still in full command mode.

Tani fumbled with his submachine gun, feeling the blood in his veins turn to ice. Moving slowly towards his companions, the young scholar gasped at his surroundings. Mineera moved forward and placed a hand over her heart at the sight of the eerie sculptures.

"What the hell is this place?" Tani quizzed.

A moment of silence passed over Nova 7. Each of them was struggling with the fear of the unknown mounting at the back of their minds.

"A tomb..." Mineera replied with an icy chill in her voice, nearly whispering. "I hear voices of the dead, angry, vengeful spirits filled with intense malice." Shaking as fright washed over her, she forced the horrid voices from her mind. Although she had regained control of her emotions, her lower lip still quivered slightly. Pulling her blue robes about her, Mineera shuddered as if to keep the chill of the tomb away from her skin. "There are hundreds of whispers rising from the shadows. In my heart, all I can feel is an icy chill. Whatever this place is, it is filled with vengeful memories of a hateful race..."

Her words were ominous and put the rest of the companions on edge. The gravity of their situation was all too real; danger was accumulating around them.

"Remember your training? Flank me, Jared on the right, Tani on the left. Stay only a step behind. If we make contact with hostiles, no one fires unless I do. Everyone got that?"

The tribals nodded in agreement. Over the last few months, Banion had begun to teach them how to fight with firearms. Both tribals of Scarskin had fired their weapons on several occasions, but were green when it came to real combat. Only Jared had ever fired a gun in battle, on the night Rasheed fell to the Reaper Kai. Neither tribal was battle-ready with a gun, and they listened to Banion's orders carefully. The surroundings of the tomb were imposing, especially considering their intent of stealing parts from the residing tribe's water purifier. Nova 7 was definitely in harm's way.

"Keep calm and fire your weapon in short bursts. Draw the barrel of the weapon downward before you fire – it will keep your weapon under control. Aim for the center of the target, don't get fancy. If we need to fight, dump as many rounds as you can into the center of whatever your target is."

The tribals nodded and felt a lump in their throats. They looked at each other with wide-eyed fright.

"Another story for our grandchildren?" Jared said to his friend from Scarskin, trying to break the tension.

Tani chuckled and felt his own spirit lift a bit.

"You sense anything else?" Banion asked, looking back at Mineera. Tension was apparent in her features, and the psychic was frantically looking back and forth, as if gripped by visions or voices. "Let me rephrase that: you sense anything alive down here?"

"No," she answered quickly. Feeling tense, she dropped her right hand down to her waist. Her dark skinned hand brushed across the hilt of her dagger. The sensation of the knife hanging from her belt gave her comfort.

Using a hand signal, Banion motioned Nova 7 into the ancient tomb.

The light from the surface above was nonexistent. As they passed through the archway, all sunlight vanished, plunging the passageway into a pitch black void filled with indecision. They were forced to rely solely on their flashlights for illumination. As the team pressed deeper into the passage, the walls of the tomb were increasingly lined with sinister pictures. Odd letters formed words in some unknown language, carved into the red sandstone wall.

Tani brushed his hand across the words as they walked. Whatever
civilization had built the stone passageway was formidable and
displayed a heightened intellect, being both wise enough to create
their own language and to build the intricate tunnels around them.

Another unsettling fact was that the inscriptions were well
worn upon the walls, showing that the passageway had been created
centuries ago. The race of creatures which built the passage had
done so in ages past. It was credible evidence that the tribe was
probably even more advanced these days, growing strong under the
earth, sheltered in the darkness for eons.

Banion led the way, deeper into the ancient crypt. His light
passed back and forth along the walls, scanning the strange passage.
A thick, sickly smell clung to the tunnels, growing gradually
stronger. All of the companions felt their stomachs lurch. The
smell of death wafted through their nostrils. Fighting to keep from
getting sick, Nova 7 started breathing through their mouths to cut
the smell. The odor was so strong, they could taste it. A grimy
stench covered them. Mixed amongst the smell of rotting flesh was
the sting of noxious chemicals, bitter and strong. The further they
pushed, the more intense the stench became. While breathing
through their mouths helped initially, it was rapidly becoming a
futile way to evade the oppressive odor. The smell clung so fiercely
to the tunnel that they could smell it in their noses even though they
refused to breathe through their nostrils. Slick mucous began to
form in their throats as bile began to rise from their stomachs.
Fighting the urge to vomit, the companions waged a desperate
struggle to simply breathe in the foreboding tunnel. They followed
Banion into the depths of the catacomb, feeling the terror rise in
their hearts.

As they reached a junction, the passageway splintered away
into several tunnels. Choosing the closest doorway, Banion led his
team into another winding catacomb of tunnels. The journey was
about to take a turn for the worse. A grisly discovery was lying in
wait. Gasping, Banion stopped dead in his tracks, paralyzed by a
terrifying sight just ahead.

In the dim light, Nova 7 could see shelves cut into the
sandstone; hundreds and hundreds of niches were carved into the red
stone walls. As Banion shone his light into one of the nooks, several
sinister forms were revealed. They were corpses, carefully draped in

a series of moldy yellow rags and stacked one on top of another in the niche. The smell of herbs, rotting flesh, and noxious chemicals filled the nostrils of Nova 7. Fighting the urge to retch, they stared at the grisly forms tucked away in the recessed spaces. The companions were chilled by the sight and smells of the mummified bodies.

As Banion trained his light on the top of the pile, a single body became the focus of attention. The form seemed to be that of a child, or rather the size of a human child. Roughly four feet in length, the corpse seemed unnaturally shaped, though the body was completely wrapped from head to toe; not a single hint of the creature's form could be seen through the grimy, desiccated, yellow rags engulfing it. The head of the creature was strange. With an elongated nose, somewhat like a snout, the wrapped body failed to divulge any additional knowledge.

"Let's get out of here," Tani called out as fright filled him.

"Yeah, forget the Slumlanders," Jared agreed.

Banion's mind was spinning; the form was familiar to him. Wild theories rolled around his mind. Nodding with a resolute expression, the leader of Nova 7 had settled on a theory, one with definite credibility.

"Maybe we won't have to help the Slumlanders in order to locate the Runner..." Banion said with a whisper. His thoughts drifted to the Consortium of Arms back in Shape Home as he surveyed the sinister body. "We might be closer to our objective than we think."

"What are you talking about?" Jared shot back, sensing some sort of ploy on Banion's end. Mineera probed him with her senses and could also sense deception coming from Banion. Whatever he was hiding, it was centered on the Runner and the strange mummified mutant corpses.

"Maybe nothing, kid. Just a hunch," the leader of Nova 7 responded, his jaw tightening as he spoke, causing the scar on his face to contort.

"What are they?" His fascination mounting, Tani crept closer, his green eyes glittering in the darkness, eyeing the strange creature resting in the niche. His curiosity was beginning to overtake his fear.

"Some sort of sickly mutant," the tribal warrior concluded,

also moving forward to get a better look.

"Look at these tunnels." Banion spun around. "No sickly mutant crafted this place. Whatever these are, we need to be extra careful."

Tani moved forward, his hand wavering above the bandages. His inquisitive nature was overcoming his proper sense of reason. He quivered with the desire to unveil the beast beneath the wrappings. Losing the battle to his own curiosity, his stumpy fingers moved forward to remove the crusty yellow bandages.

"No!" Mineera warned Tani. She shot forward and grabbed his hand, pulling him away from the mutant corpse. "It's not wise to disturb the dead. Let these foul beasts rest."

Tani's hand shot back. Fear overrode curiosity once more as Mineera startled him. Abandoning his inquiry, he retreated away from the corpse.

Panning their lights around, Nova 7 experienced a feeling of awe washing over them. The tunnels and passages stretched on in every direction, beyond their field of vision. A dense series of passages and catacombs snaked away into the earth. Each passage was filled floor-to-ceiling with niches cut into the walls at regular intervals. Each niche was filled with more mummified corpses of the strange mutant species.

"There must be thousands," Jared stuttered out.

"Tens of thousands," the tribal scholar amended.

"Generation upon generation of dead mutants are down here." Banion spoke in a haunted tone, fighting back the smell by exhaling the words strongly. "Let's get out of here as soon as possible."

"Where do we go now?" Mineera asked, mystified, and with good reason. Passages were everywhere, leading off into the darkness.

Scanning his crude map given to him by the Slumlander Prefect, Banion shook his head with indecision. "I have no idea where we are..." The map was useless, and failed to divulge any of the tunnels' secrets.

Sighing, Banion surveyed what lay ahead: a twisting mass of passages dug under the earth, filled with desiccated bodies. Taking a hesitant step forward, he chose one lonely passage.

Nova 7 crept deeper into the eerie catacombs. The further

they pressed, the further they moved from safety. It didn't take long for the mercenary team to get hopelessly lost in the maze of twisting rooms and tunnels filled with the mummified corpses of the dead. Their anxiety and indecision grew as they passed from tunnel to tunnel, room to room. Each would occasionally look into the shelves with quick, fleeting glances. The remains of the dead were taunting the heroes with their lifeless silence, enveloped by the crusty yellow rags. Moments turned to minutes, and minutes into hours as the team members walked amongst the dead. In the shadows of the passage, a whisper rose as the rage of the corpses worked its haunting magic of fear upon the companions. If Nova 7 didn't find an exit soon, the chance that the yet-unknown living would find them here, amongst the dead, would grow to be inevitable.

Chapter 11
Fields of Gold

It had taken the mercenary team several hours to liberate themselves from the twisting catacombs filled with thousands of mummified mutant corpses. Finally, the horrid smell of rotting flesh had abated. Breathing clean air once more, Nova 7 heaved a sigh of relief.

The companions were standing near a wide circular passageway of broken concrete. With the memory of the foul smelling crypt growing more distant, they gazed ahead to the circular pipe of concrete.

The earth had fractured this portion of the underground tunnel. An ancient sewer pipe had broken just beyond the edge of the crypt, allowing movement north and south along the old passage. Standing at the junction, the companions eyed one another with a look of concern. It had taken hours to escape the catacombs, and now another choice presented itself.

Gesturing in frustration at the broken sewer lines, Banion was about to concede defeat. "Next time, I'm listening to the runts," he muttered with a hint of disgust, eyeing Mineera. "We should have never accepted this insane quest."

"Which insane quest?" Tani mused, trying to get a rise out of Banion. "Stealing water purifier parts, or locating a nuclear weapon for a kingdom that got wiped off the face of the earth in one night?"

Ignoring the tribal, Banion quizzed the others for a course of action. "You sense anything, Mineera?"

Her blue eyes stared straight ahead, her expression distant. As she meditated from her stationary position, her powers stretched outward, trying to find a scent of life and discern the next course of action for Nova 7. Despite her concentration, she remained blank, unable to read anything from their surroundings. "I sense nothing," she replied.

Sighing, Banion stood at the junction and spun around, his indecision mounting.

"Let's head south." Jared spoke in a confident tone, holding a crude compass in his hand as he stared at the junction.

"South?" Tani asked, scrunching his nose. Puzzled, he pushed his wire-rim glasses up his nose and stared at the tribal warrior from Scarskin.

"Yeah…" Crouching, Jared ran his hand over the ground at the junction. To the south, the earth was worn, as if the passing of many had eroded the stone over the centuries. The northern passage was mostly undisturbed. Rough rock still clung to the floor. "The southern passage is well worn. If I had to make a choice, I would head south."

"That's good enough for me," Banion concluded, taking a step forward toward the broken sewage pipe.

Lighting the way with the flashlight still mounted on his assault rifle, the veteran mercenary moved forward slowly, eyeing the ancient sewer pipe with disgust. The journey through the underworld had not been pleasant thus far, and the thought of traveling in a sewer pipe was not a pleasant one.

The beam of the flashlight scanned the floor of the round concrete pipe. The sewage had long since evaporated, and a thin layer of dirt littered the bottom of the pipe. Strange, inhuman tracks were pressed into the sediment upon the floor, further evidence of the mutant tribe inhabiting the underworld labyrinth beneath the Concrete Barrens.

The rest of Nova 7 eyed the tracks with concern. If they kept pressing deeper into the ruins, contact with the inhabitants would be inevitable. Taking the plunge, the team moved into the sewage pipe and traveled south into the ruined world beyond.

Crouching, the team members moved along the concrete drain, bending down to avoid hitting their heads on the ceiling of the pipe. The darkness of the pipe was cloying, seeming to hover right

on top of them. Each of them could hear their own breath echoing
in the tight, claustrophobic pipe.

With every step they took, the walls seemed to be pressing
ever closer. Several times, Jared scraped his head on the top of the
pipe. Tani felt an odd sense of foreboding. It seemed to him that
the pipe was narrowing, or maybe… it was just his imagination. As
the confined concrete tube wore at their senses, a hint of salvation
loomed near. At the end of the pipe, a glow could be seen; light, a
bright white pinnacle, was just ahead, bathing the end of the tunnel
in warm hues. It seemed like an illusion at first, but grew
progressively more intense as the team moved forward. Soon, the
light source was so bright that they no longer required their
flashlights.

The warm glow of the room beyond filled each member of
Nova 7 with a sense of awe and wonder. How could there be, so far
beneath the earth, such a radiant source of light? Befuddled by the
oddness of the warm light, the group sought to discover the mystery
of the underground passage. Brandishing their weapons, they
moved out of the ancient sewer pipe into the expansive room
beyond.

To say Nova 7 was stunned by what they found was an
understatement. Ahead of the team were fields, golden fields of
wheat, tucked beneath the earth, bathed in the warmth of white light.
Scanning the room in confusion, the companions quickly discovered
the source of the warm glowing rays. The western wall of the cave
had eroded, allowing sunlight to enter the cave. Brilliant rays
poured into the cavern. Squinting in the bright daylight, Tani
gasped and moved into the fields. Holding out his hands, the young
scholar felt the tall stalks of wheat brush against his skin.
Meandering through the golden field, secluded in the cleft of the
earth, the scholar was lost in his own thoughts.

Jared moved after him, advancing toward the collapsed wall
on the western edge of the chamber. To his surprise, the edge of the
cavern was enormous, as if an earthquake had shattered the wall in
ages past. Reaching the edge, the tribal found that he was standing
at the top of an enormous cliff, high above the violent ocean below.
Staring downward, Jared found himself more than a hundred feet
above the violent surging surf. Waves crashed against the rock face
with powerful sprays of white sea-foam. Enormous blue surges of

water rolled forward, relentlessly slamming the broken edge of the city. Turning around, the tribal warrior motioned the others over to his side.

Feeling the warmth of the sun upon his skin, Jared eyed the ocean with wide eyed wonder. The surging sea beyond the jagged cliff face was dotted with a collection of ruins, most barely visible above the ocean waves. In ages past, the city had splintered and fractured under the powerful will of the violent earth. A portion of the city had collapsed into the ocean, leaving only the remains of tall buildings peeking out from the wild waves. Also littering the violent surf was a host of small, earthen islands dotting the ocean, each isle covered by more ruins. For miles and miles, small portions of shattered earth were fully exposed, held above the waves with buildings of the ancient world still mostly intact. The chain of islands stretched north along the coast, the furthest of them disappearing into a low-hanging bank of fog.

The team viewed the broken city below with a dreamy gaze.

"The Shattered Isles..." Mineera said, her voice haunted, staring at the small islands dotting the ocean.

"What?" Banion inquired.

"The Shattered Isles. It's an ancient Reaper Kai legend. Scouts of my former empire in ages past claimed to have reached the western coast of the continent. What they found was called the Shattered Isles, portions of this ancient city that still remain above the ocean waves. I thought it was just a legend, but apparently not." The traitor of the Reaper Kai empire spoke with a misty look in her blue eyes.

"Legend no more," Jared said, smiling. "No one from Scarskin will ever believe the story of this place." The islands forming the chain were huge, stretching many miles across the ocean. After surveying the vast expanse before him, Jared felt more confident about their decision to steal the water purifier parts instead of trying to scour the countless islands in the ocean below. Trying to locate the purifier parts amongst the ocean ruins would have been *nearly* impossible.

"Amazing," Tani murmured in a dreamy tone, still surveying the wheat field growing strong in the rock cleft, protected from the contaminated world of the radioactive city above, sheltered from the wild ocean waves crashing against the cliffs below. The wheat grew in rows, and looked well tended. Tons upon tons of soil had been painstakingly placed in the cave, providing the crop with a nutrient base. Whoever – or whatever – grew the fields was well suited to agriculture. The mutant tribe beneath the earth was formidable indeed. His green eyes darted back and forth. "These fields are large enough to supply food to thousands of inhabitants, maybe more, judging by the size of their dead we found in the catacombs." Tani looked over at Banion.

Banion looked back and their eyes locked. Though amazed with their surroundings, the leader of Nova 7 seemed to almost *expect* what they were seeing.

Tani got the impression that the mercenary was hiding something, almost as if he knew something more about the catacombs, the wheat fields, and the strange mutants controlling the tunnels they had been traversing. Banion averted his gaze, keeping his secrets to himself.

"Let's get moving. I think we are nearing the end of this little adventure," Banion said, ordering the rest of the team to press onward. With one final look, Nova 7 gazed upon the eerie golden field of wheat, bathed in the warmth of the setting sun. Leaving the calm, serene peace of the fields, the team pushed into the southern edge of the cavern and found another tunnel moving into the shadowy darkness beyond.

It took a mere fifty feet of travel from the fields of wheat for the team to come across another cryptic sight. Just ahead, the passageway ended at a set of enormous steel doors, rusted and corroded by the passage of time, covered by strange text in some unknown language. Two enormous iron handles were set in the door, only a foot or so off the ground. Whatever was accustomed to opening the doors was small in stature.

Banion brushed his hand across the corroded door and hesitated. "This is it. I can feel it," he said in a resolute tone.

Nodding in agreement, the team prepared to face anything that might erupt from the unknown. Drawing his silver, long barreled revolver, Frank's revolver, Banion held the gun ready as his other hand strained against the iron handle. The door shuddered and the hinges groaned under the force of his strength. As the door opened, a rush of humid, sickening air exploded outward from the opening with a sinister hiss, as if a seal had broken. The smell of dirty, wet animal rolled outwards. It was a nasty scent, reminding the team members of filthy dogs and excrement.

"Banion!" Mineera cried out in distress. Something at the edge of her senses was tearing at her mind, a sinister warning of what lay beyond. "I feel great evil from within."

Jared steadied his submachine gun against his shoulder, as did the scholar from Scarskin.

"Look sharp, everyone. Watch your backs." As they moved inside the doorway, the smell of dirty animal wafted more strongly through their nostrils. Whatever the source of the smell, the companions felt as if they were heading into an animal den.

The floor was moist and a revolting, grimy fungus covered the floor. A moldy smell erupted from the passage. Slimy, brown mold covered every surface in the cavern. The tunnels were carved within the rock, small, circular passages barely the size of a full-grown human. Spheres were inserted in the rock at regular intervals, emanating a dim, grungy yellow light. A cable running along the wall linked all the lights.

"Light bulbs?" Tani mused, eyeing the strange spheres lining the walls. His suspicion was confirmed as he touched the electrical cables connecting all of the lights. The scholar was stunned by his discovery. It just didn't seem possible to find such advanced technology still existing within a city that had long ago been ravaged by nuclear fire. Shaking off the distraction, Tani hastily moved after his companions, who had pressed on down the passageway.

Following the twisting tunnel, the passage moved deeper into the eerie underworld. In the dim yellow light, the shadows seemed to heave, as if something lingered at the edge of their vision. Steadying their weapons, they moved on.

The sound of whining metal rolled through the passage. Somewhere up ahead, machines were grinding away. The lights flickered along the wall, flashing with a disturbing rhythm.

A gigantic room opened up ahead of them. A waterfall cascaded from a hole in the roof, fed by a water source from the contaminated city above. Grinding and pumping, an enormous waterwheel rotated under the force of the waterfall. Rusted steel wedges captured the falling water, growing heavier and allowing the mighty wheel to rotate. Turbines turned with each pass of the wheel, creating a flow of electricity. A host of thick cables erupted from the groaning turbines, sending an electrical current into open conduits in the wall.

The mighty waterwheel was stationed above a pump house. At the base of the pump house was a foul-smelling pool, churning with stagnant water, so dark it was nearly black. The contaminated water was being sucked into the pump house and out the other side into an enormous holding tank. Nova 7 had found their target, the pumping house that held the purifier parts they needed to seal a deal with the Slumlanders.

"There it is," Banion whispered, pointing at the pumping station. "Let's get down there and get what we need. The water purifier must be in that pump house."

"I can feel many whispers here," Mineera warned. "I can't tell where they are, but many sentient beings are close by. We must be on our guard."

"I have been on my guard ever since we climbed into that damn hole," Tani replied, his eyes wide.

"Get it together, bookworm. It's your time to shine and get those damn purifier parts."

The group moved quickly down the slope towards the pumping station. No living beings were evident as they advanced. The pathway seemed clear. Charging forward, they reached a metal catwalk spanning the murky, black pool of stagnant water and leading to the pumping station. Bounding across, Nova 7 made it midway across the catwalk before they caught sight of the enemy.

A hiss rose from the darkness, emanating from an indistinct flurry of movement. Small forms, the size of human children, were crawling across the very walls of the room. Small furry bodies, a mixture of gray, white, and brown, clambered across the walls. Like a swarm they came, a host of wild vermin dressed in leather clothes: mutant rats, all brandishing firearms. Walking on their hind legs, these mutant rats had humanoid hands, each clutching a weapon of some sort. The sudden burst of movement was so quick, it caught all of Nova 7 off guard.

Crouching upon the catwalk over the wretched pond of murky water, the team prepared for the worst and took up defensive positions.

"Damn it!" Banion shouted. "We need to get out of here!"

It was too late. Within a matter of seconds, hundreds of mutant rat warriors had surrounded them, training their weapons upon them. As they hissed like wild beasts, the black eyes of the mutant rats glared at the humans with hatred. Their hair standing on end and their hunched posture aggressive, the vermin were fully ready for conflict.

Each side of the catwalk was covered with enemies, foul smelling vermin cutting off all hope of escape. Their force was more than two dozen thick, with additional reinforcements now covering the roof of the pump house.

With a horde of mutant rats surrounding them, the mercenary team was out of options. There were too many to fight, and escape was impossible. These grim circumstances were the result of the naïve mercenary team's invasion of the underground ruins. If something didn't happen fast, some sort of wild miracle, the rabble of vermin would slaughter the hapless mercenary team. As indecision and fear overwhelmed the companions, the rats moved slowly forward across the bridge. The black, angry eyes of the mutant rats swelled with hunger while they hissed with sinister fury. Their hate was about to be unleashed, and Nova 7 was the target.

As the host of vermin edged forward, the team looked on in horror. The frantic companions braced themselves for attack, mentally preparing for the worst.

Chapter 12
The City of Verminhold

As it opened its mouth, the jagged yellow teeth of the mutant rat gnashed together, creating a grinding sound. With patches of white fur on its mostly jet black body, the mutant rat was an imposing sight as it stood on its hind legs like a human, a shotgun clutched in its clawed hands. Growling, the mutant rat hissed while pointing its firearm at Banion.

With a grim look, Banion held his revolver ready. If he was about to die, he would make sure to take as many mutant rats as he could with him. Jared and Tani were paralyzed with fright. Mineera was meditating, surrounding the group in a barrier of psychic energy; she was strong, but holding the psychic shield around them all was taking a heavy, taxing toll upon her soul. She could not protect them forever.

Dozens upon dozens more of the mutants backed their leader, clambering around the pump house. In a matter of seconds, hundreds of mutant rats were everywhere, many of them clinging to the rock walls of the enormous cavern.

"What matter of business brings foolish humans into our domain?" the rat warrior hissed, still brandishing its shotgun.

Banion's eyes narrowed as he stared at the mutant warriors. It all made sense. The strange cloaked forms in the Consortium of Arms many moons ago had looked oddly like the creatures standing before him now. Taking a gamble, Banion wagered all of their lives on an intangible hope. "I am Banion O'Neil of Nova 7."

Banion's response rolled through the rat warriors with

startling results. Upon hearing his name, the collective of rats began to hiss in glee. The sound of hundreds of voices growling in unison was a haunting one. A greedy, mischievous feeling emanated from the mutants; they were obviously elated by Banion's answer. Lowering his weapon, the jet black rat smiled and moved towards Banion, its hunched form gliding forward quickly. The other members of Nova 7 were horrified by the mutants' reaction. It was as if they had perceived Banion as some sort of god.

"At last..." the rat warrior hissed in glee. "Our champion has arrived!"

"Champion?" Jared echoed, hostility flaring in his eyes. "What have you done?" the tribal warrior asked Banion, confusion, terror and a growing anger all flashing upon his face.

As the rats moved forward, Mineera's concentration was spent; the psychic shield dropped, and a wash of dizziness flooded her. The sinister feeling of evil invaded her senses. The creatures moving towards them viewed Banion as some sort of savior, and this put Mineera on edge. Whatever Banion had done, the mutant rats were *his* allies; a twisted thought indeed.

"Our savior!" another rat hissed, moving forward as its black eyes stared at Banion. With furry arm outstretched, the mutant moved to touch Banion.

"Banion, what the hell is going on?" Tani asked, his expression similarly horrified.

The leader of the rat warriors stepped toward Banion and bowed before him. "The Runner will be most pleased that you have finally reached our city."

"The Runner?" Jared said, feeling betrayed. It seemed that Banion knew much more about the Runner than he had led the others to believe.

"Yes... our leader is the Runner... King Lavosi, the Vermin Lord!" a rat warrior hissed from behind them. "Come with us! We will lead you to him!"

Mineera, Jared, and Tani all shot Banion looks of concern. They all felt betrayed. What secret dealings had Banion conducted with the sinister mutant tribe? Of what plot were they all now a part? These and other questions darted frantically through their minds. Whatever had happened, all of them felt that a great trust had been breached. Banion ignored their stares; instead, he simply looked ahead, toward their objective, with an intentionally blank expression.

The tribe of vermin led Nova 7 into the hollowed reaches of the underworld, ever closer to the mighty city of the mutant rats.

As they moved beyond the pumping station, the cavern opened into an enormous chamber. Rising several hundred feet in the air, the walls of the chamber disappeared into the darkness above. The sight greeting them filled Nova 7 with shock.

Built upon the walls of the mighty chamber were small platforms, bolted to the cavern walls and supported with strong steel cables. Upon each platform were more mutant rats, staring down with hungry eyes, watching the newcomers with sinister intent. Each perch housed a collection of rags, garbage, and debris. The foul nests were writhing with activity as young rats skittered about amongst the refuse. The mighty chamber was home to thousands of sentient vermin. Cargo netting covered the walls between the platforms, allowing the host of vermin to move freely upon the walls of the enormous cavern.

As the news of visitors rolled through the rat collective, the vermin clambered to the walls, climbing down the cargo netting. Soon the walls were covered with thousands of rats, moving swiftly downwards to view the newcomers to the city. A sickly smell coursed through the room, an odor of feces and dirty animal mingling sickeningly.

"Behold!" the rat warrior roared in glee. "The mighty city of Verminhold!"

Spinning around, the team members felt their skin crawl. Thousands of sinister rats were all around them, on the ground, on the walls, everywhere, staring down at them. Forcing claustrophobic thoughts and fear from their minds, Nova 7 continued onward, following the rat leader deeper into the mighty city. As they left the 'housing district,' another series of caverns opened before them.

The next series of caves displayed scenes of industry. Mighty foundries had been built; a wondrous collection of metal works were suspended over a fiery pit. A pool of bubbling lava lit the room with an orange glow. Above the fiery chasm, upon a steel catwalk, rat crafters were melting steel. Using the heat of the magma to melt the metal, the vermin metal workers poured the liquid steel into molds to create a variety of tools. The foundries were large, and several dozen workers went about their business intently, crafting tools and weapons of war.

Viewing the ingenuity of this intricate civilization, the members of the mercenary team were in awe. It was surprising to find so much activity and technology deep beneath the surface of a city ravaged by nuclear fire in ages past.

Beyond the foundries was another set of workshops. Using the tar found all over the volcanic region, the rat tribe had developed the art of extracting chemicals from the tar. With crude cloth filters over their faces, more rats labored over bubbling vats of tar, stirring the nasty brew. Primitive extraction funnels rested above the bubbling tar, capturing the volatile vapors seeping from the black ooze. The funnels spiraled upwards and near the top, vapors coalesced and cooled. Drops of clear liquid emerged from the other end of the funnels, collecting in steel barrels. The rats were brilliant indeed, as evident by their success in learning how to separate and purify the volatile contents of the tar. Tani was in awe. The potential for creating and synthesizing chemicals was possible. The advantages of chemical processing of this sort were staggering.

"Impressive, yes?" the rat hissed at Tani, seeing the wonder upon his face.

"Yes, it is," Tani concurred. "Formidable indeed."

Continuing their journey, the group moved into another cavern filled with technological wonders. The next series of passages were home to an imaginative collection of machines. Giant looms, over two stories tall, had been constructed. Using a steam vent, the vaporized water had been channeled into pipes which ran to the top of the looms. Powered by steam, the looms worked feverously to manufacture an array of fibers. Cloth was being formed on one such loom, driven by the steam power harnessed from the earth. Large bundles of fabric were being collected at its base by a horde of rat workers. Other machines were using the material to shape swaths of polymer-based plastics, using the raw materials harnessed from the toxic tar vats. At the far end of the enormous room, a mighty device was slowly churning out steel cabling, suitable for conducting electricity.

"They have us beat," Jared whispered to Tani. The young scholar agreed, creating mental pictures of the devices around him.

"Scarskin is a primitive hovel compared to this technology."

"After seeing this, I am positive we could construct similar devices back in Scarskin. We could use the steam vents in the sulfur caves to power our own machines." The scholar's wild eyes, and his hungry mind, were devouring the wondrous machines. The Exile was of true benefit. Having seen advanced technology, the tribals could radically change the village of Scarskin upon their return.

"The elders were wise to exile its youth into the wasteland. When we make it home, we can advance our village to this level of technology or beyond," Jared said resolutely.

"I never imagined such a place." Mineera shook her head, overhearing the conversation of the tribals. "This is truly amazing."

Hearing the companions speak of their city in such a manner filled the rat warrior with pride. "We are strong!" it hissed in glee.

Coming to the end of their journey, the companions moved into an expansive tunnel. In ancient times, the tunnel had housed a subway station, which was now a mere hint of its past glory. The tunnel had collapsed at the rear of the passage, sealing it from the underpasses beyond. In the center of the room, bathed in the yellow glow of primitive light bulbs, was an enormous throne, built upon a finely carved slab of tan sandstone. The throne was covered with gold, silver, and other precious metals, and surrounded by a host of guards, all dressed in sophisticated, old-world body armor. Thick bulletproof vests with steel trauma plates beneath the armor gave the mutant rats wearing them a sturdy, solid appearance. Far from being obese, the rat warriors were strong, and their muscles rippled under their fur. Each guard was heavily armed, brandishing automatic weapons almost as large as the creatures holding them.

The occupant of the throne was an odd sight. Sitting upon the gold inlaid chair was a form garbed in finely crafted brown robes. Thick leather gloves covered the hands of the creature, and a hood was drawn over its face. Banion knew the creature sitting upon the throne. It was the same mutant he had encountered long ago, in the Consortium of Arms, the same creature with whom he had struck a deal, promising to be the champion of his tribe.

"Banion O'Neil," the creature growled. "I am glad you have finally reached our home."

Banion nodded, paying his respects to the creature upon the

throne.

"Let me formerly introduce myself." Drawing back the hood from its head, the king of the rat tribe exposed its face. Pale white fur covered his head and face, a stark contrast to the brown robes covering the rest of his body. His nose was bright pink and his teeth were an ivory white. The rat king smiled as his fiery, blood-red eyes stared at Banion, and then shifted his gaze to the rest of the group with sinister intent. All of them felt a wash of evil emanating from the albino rat king; you did not have to be psychic to *feel* the malice and greed coming from him. "I am Lavosi, the Runner, King of Verminhold."

Standing before the Runner filled the members of Nova 7 with a sense of closure. For many months, locating this arms dealer had been the focus of their journey. The plot to steal the water purifier parts for the Slumlanders was now a distant memory. No longer did Nova 7 have to steal the parts to locate the Runner. By a strange twist of fate, they had found the Runner on their own.

"Good thing we didn't steal from these *things*," Jared whispered while eyeing the mutants. Mineera and Tani both heard him and nodded in agreement. It would have been certain death to steal the purifier parts form the rat tribe, not to mention ruining any chance of locating a nuclear weapon. It was a fortunate turn of events.

"This is Nova 7. Tani, the scholar. Jared, the tribal warrior. Mineera… traitor and former diplomat of the Reaper Kai." Banion introduced the rest of his team to Lavosi.

"Ah, Reaper Kai!" the rat king spoke, turning his blood-red eyes towards Mineera. "So wonderful this world is. To reach your goal of slaughter and vengeance, the mighty Banion O'Neil travels with the very enemy he seeks to destroy. Tell me, Mineera of the Reaper Kai, what is it like to travel with someone who is obsessed with the slaughter of your race?"

Mineera stared back, her bright blue eyes luminous against her dark skin, and focused upon the blood red eyes of Lavosi amidst the backdrop of his ivory fur. She considered him with a look of dispassion; Mineera knew better than to give up any ground to this twisted mutant. "The Reaper Kai are no longer my kind. I also seek to rid the world of the Dark Order." Her tone was emotionless, and the rat king smiled at her response. She was undaunted by his taunts and he saw strength in her.

"Oh, and look, tribals!" Lavosi hissed, taunting the youths, his crimson eyes shifting to Tani and Jared. "And so young…"

Jared shot the vermin king a look of hate while his tribal friend simply averted his eyes.

"You have some spunk, don't you?" Hissing at Jared, the rat king smiled. "You will all do nicely," he concluded.

"Let's speak about our deal," Banion said in a harsh tone.

"Oh, our deal? Of course." The albino rat king smiled, his lids narrowed around his blood red eyes. The diabolic stare emanating from Lavosi made him look like a sadistic lunatic. "The Slumlander tribe that inhabits the upper reaches of this ruined city has become a thorn in my side as of late. Their water purifier has failed, and they seek to destroy my tribe to steal the parts from our water purifier. Though they lack any real technology, they are strong in number, very strong in number. We are waging a losing battle against the Slumlanders. They attack frequently, and even though we have strong weapons, they are destroying us slowly. Soon we will not be able to survive their advance and we will be slaughtered."

"Our original plan was simple. The city is already contaminated by radiation from the apocalypse. We managed to locate a nuclear warhead and were going to use it on the Slumlanders. But our expedition to liberate the nuclear weapon from the ruined military base failed. We were unable to free it from the vault it is held in…" His speech fell silent, his expression growing distant. Lavosi was hiding *something*.

"So, what is the deal?" Banion pressed on, excited by Lavosi's disclosure that they had located a nuclear warhead previously during their exploits.

"The plan is simple. You are to *ally* yourselves with the Slumlanders, gain their confidence, then report back here, to me. For many months, we have been extracting toxins and poison from the tar pits in the Concrete Barrens. After you have befriended the Slumlanders, you are to take this potent poison and taint their water supply."

"What?" Tani said in alarm. "You want us to kill the Slumlanders by poisoning them?"

"Yes!" Lavosi hissed, his blood red eyes bulging in anger.

"Even though their water supply is tainted already, they will eradicate us before they slowly die from their contaminated water. All I want to do is accelerate the process, kill them quickly!"

"What if we refuse your *deal?*" Jared shot back, flustered by the request to murder thousands of humans.

"You cannot refuse this deal! Banion has already sealed a pact with us, back in Shape Home. All of your gear, guns, and equipment were provided by me. It is too late to refuse. You will kill the Slumlanders, and in return for your deeds, the location of the nuclear warhead will be yours!" the rat king hissed harshly at Jared.

"What have you done to us?" Mineera's expression was horrified as she stared at Banion. "Is your quest for vengeance so strong that you would slaughter anything in your path to find retribution?"

Banion's eyes were brimming with aggression. "The Slumlanders are dead when the Reaper Kai reach the Concrete Barrens! If it takes the sacrifice of the weak, people who cannot hope to stand against the Reaper Kai, then so be it! All we are doing is delaying the Slumlanders' death!"

"Ironic, isn't it?" Lavosi smiled. "The Reaper Kai witch is protecting the weak, while the hero of Dune Station is ready to kill anything in his path to satiate his hunger! How pleasant is this exchange?"

"Banion, why the hell would you cut a deal like this?" Tani's cheeks flushed red with mounting anger. "I am not a murderer."

"Shut your damn mouth, bookworm! I didn't know *all* of the details before I sealed this deal." Banion was fuming, and didn't seem like himself. Blind anger had taken over.

"Would it have mattered? Huh, Banion? Would the details even have mattered to you?" Jared defended Tani, anger flaring in his eyes. "Is your twisted sense of vengeance so corrupt that you are willing to commit murder?"

"It's not just murder. This is an act of genocide, destroying an entire race of people by a savage act of hate!" Mineera's tone was vicious. "I did not betray the Reaper Kai order and struggle against the darkness only to join another faction with murderous intent. I do not want any part of this."

Banion fell silent. He had crossed the line, making a deal

with a sinister mutant without knowing the repercussions of his actions. The Runner had played Banion, and the gunfighter had fallen into the trap. In order to save the free peoples of the Darken Realm, Nova 7 would now have to make an impossible choice: murder a tribe of humans in exchange for a nuclear warhead, or allow the Reaper Kai to destroy the entire Darken Realm. In this world of madness, Nova 7 would have to murder the innocent to save the lives of hundreds of thousands more.

As silence gripped the team, Lavosi, the dark king of the underworld city, smiled in glee. The dissention he had caused within the members of the mercenary team was tearing them apart. As his blood red eyes washed over each of them, he knew that the fresh sense of betrayal driving them apart would make it very easy for him to manipulate them. Knowing that he had control over them, Lavosi smiled once more. In dealing with the moral dilemma thrust upon them, Nova 7 was now in a very dark place indeed.

Chapter 13
Mission of Madness

A calm came over the ruins as night broke, driving away lingering shadows with imposing darkness. Resting at a height of over ten stories, in the remains of an ancient office building, the members of Nova 7 brooded, with indecision and dissention running high. The concrete supports were still in good condition, with the exception of the northeastern corner of the building, where several stories had collapsed ages ago. Old tarnished chairs and battered desks littered each floor. Smashed computer monitors and wreckage from the shattered glass windows added to the layers of ancient garbage clogging the top floors. Anything of value had been ransacked hundreds of years ago, leaving only litter.

Tani was sitting on the edge of a window sill. His legs were hanging over the side, dangling in the air. Jared had taken up a position next to his friend from Scarskin, his back resting against a concrete support column at the corner of the room, his eyes turned west, toward the horizon. Lost in thought, the two tribals gazed with distant stares as night began to fall over the ruins once more.

The sun was beginning to push westward, beyond the edge of the world. A bright explosion of color lit the darkening sky. Fiery orange, blazing reds, and soothing purples lit the clouds, painting a vibrant picture over the blunt and jagged buildings and lonely byways of the Concrete Barrens.

As the sun set, a mixture of emotions was gripping the tribals from Scarskin. Flooded with feelings of mistrust, they wondered how they could have been so blind for so long. They both felt as if they were nothing more than pawns in some sort of twisted chess game, with Banion controlling all of the pieces. As the reality of what Banion had done to them began to sink in, both Jared and Tani wondered how they could move on and finish the insane crusade to stop the dread Reaper Kai. Their emotions began to boil to the surface as their frustration mounted. Sighing with irritation, Jared grabbed his ponytail, pondering for a brief moment if he should rip it from his scalp in an act of utter despair.

"I think we should leave. I don't want any part of this crap. I am absolutely sick of this. I'm sick of Mineera going crazy. I'm sick of Banion taking us for granted and manipulating us. And most of all..." Jared rose to his feet, flailing his arms about with his voice turning to a shout, "I'm sick of being in the middle of nowhere!"

Trying to further his point, he cupped his hands around his mouth as if to amplify his rants. Breathing in a lungful of air, he shouted once more. "Hello? Is anybody out there?" With a dramatic gesture, he then cupped his hand around his right ear as if to amplify any possible response from the battered city beyond. Only silence responded.

"Are you finished?" Tani barked back snidely. "Nothing lives out here."

"Damn the Slumlanders and damn those rats!" Jared said, his tone still heated. "I don't want any part of their war."

Tani looked back at Jared, continuing to kick his dangling legs back and forth over the edge of the window sill. With a sigh, he shook his head in frustration, agreeing with his friend. "I am no murderer. I don't know what to do."

"This isn't the first time something like this has happened. Remember Rasheed? Remember walking into that war room and finding out we had been enlisted into Banion's team? Banion betrayed us back then by manipulating us, and he has done it again."

"Yeah, except this time, the outcome of the manipulation is pure evil, a hideous act of violence to satiate Banion's sick hunger for vengeance. Our actions are unconscionable if we accept this mission. Killing thousands of humans to get a *chance* to locate a nuclear weapon? What kind of deal is that?" Tani shook his head as if attempting to distance himself from the situation.

"It's not a choice." Mineera emerged from the growing shadows, pausing to rest before the tribals. Unceremoniously, she sat down and joined their conversation. With her blue eyes growing darker by the second in the dimming light, she considered the tribals, a look of distress upon her face. "I have been mulling over this decision the entire day. In the end, without the nuclear warhead, the free peoples are defenseless. Without the power of the ancients, the Darken Realm will fall to the Reaper Kai."

"If we murder the innocent, does it matter who wins? If we kill needlessly to achieve our goal, what's the point?" Jared asked, gazing intently at Mineera.

"If we become monsters, there is no winner in this conflict," Tani agreed.

"Such a decision is a heavy burden to carry," Mineera concluded. "It's not an option, in my mind, to needlessly slaughter the innocent in order to protect the greater good. Perhaps if I were still under the rule of shadow, such a decision would be easy. But to abandon our hearts and souls for savage hate is a thing that I will not be a part of. You are correct, Tani: to act as monsters to stop other monsters is a losing proposition." Her words were profound, and they caused the trio to abandon their conversation, each seeking to further ponder the situation on their own.

All three companions were lost in thought. What was the right thing to do? Allowing the Reaper Kai a reprieve by abandoning the quest to find a nuclear weapon would mean the destruction of the Darken Realm. On the other hand, killing the Slumlander tribe would be a horrid deed, something unconscionable, but would save many more lives and put an end to the dread Reaper Kai. Sacrificing the weak to save the world was a tricky matter. And what of moral consequence? Was destroying an enemy worth losing the humanity of one's spirit? Was murdering a few thousands justified by saving hundreds of thousands more? It was a grim choice, with no clear solution.

The fiery orange shifted to red, then purple. As the last hues of fading color darkened and turned to night, the trio sat in the solitude of the ruined building, staring out at the ominous site of the battered city. Nothing stirred amongst the fallen concrete and corroded automobiles. The city was quiet as the stars erupted from the veiled darkness. The stunned silence that gripped them was eerie; each of them was submerged in inner turmoil, coming to grips with what they had to do. In the end, the decision was horrible, but each of them was coming closer to making the inevitable decision: sacrifice the weak to slaughter the strong.

He arrived like a wraith, slinking from the darkness of the office building like smoke drifting through the night. Dressed in his long overcoat, Banion closed in on the rest of Nova 7 without making a sound, coming within a few feet of them before any of them noticed. The darkness of the night made the right side of Banion's face, scarred from past violence, look like a tortured growth, a sinister mass of memories and sorrow. His dark eyes seeming to glitter in the night, he came before them to confront each of them. The mad hero of Dune Station spoke in a direct, assertive tone, his posture aggressive. "I have come to a decision." Pausing, Banion surveyed the rest of his team, hoping for a response.

Each of them looked away from him as he spoke. The reality of what they had to do made each of them hate him, hate him for being the voice of reason in such vile times. Not wanting to concede the wisdom behind the decision of sacrificing the weak Slumlander tribe to obtain a nuclear weapon, each of the team members sought to bury their own emotions.

"I have been out here for too damn long, so long I can't even remember my way home. All of my life, bloodshed and violence have ruled my existence. All I know is how to reach my objective, at any cost. Somewhere, back there, in the ruins of this once mighty land, I lost my very soul." Shaking his head, he began to laugh dryly, wishing his pain and suffering away with a shaky display of mirth.

"I got so caught up in my obsession for vengeance, I forgot why I ever came out here. For the last few hours, I have tried to rationalize the decision I made. In that time, I came to realize my Lily, my sweet Lily, would never smile at what I have become. My uncle and my mother would also be ashamed of me. Hell, I am ashamed of me." Sighing, Banion rubbed his head with his hands as if trying to force the gravity of the situation from his mind.

"I have killed and shed the blood of the innocent. I will never do it again. From each of you, I have learned that ideals and compassion are the way to win this war, not reckless hate."

Each of them turned to view Banion. There was a quiver in his voice as he spoke. Fighting back the tears, he smiled and closed his eyes. Breathing in resolutely, Banion kept his emotions in check.

"I am proud of each of you. Tani for his knowledge, Jared for his courage, Mineera for her spirit, and each of you for your heart. You have all taught me more than you could ever know." As he smiled with tears in his eyes, each of his companions looked back and fought the surge of emotion welling up inside their own chests.

Not knowing what had brought about this speech, Tani, Jared, and Mineera dared not speak.

"We are not going to kill the Slumlander tribe; instead, we are going to steal a boat," Banion concluded with a glimmer of madness in his eyes.

A stunned silence rolled over Nova 7. The trio stared in horror at the smiling Banion. Was this finally it? Was this finally the moment when Banion totally lost whatever remained of his fragile sanity?

"What!?" Tani cried out, exasperated, scrunching up his nose as he stared Banion down through his wire-rim glasses.

"What the hell are you talking about?" Jared added, crossing his arms across his chest. Rising to his full height, the disturbed tribal warrior was clearly on edge.

"Steal a boat?" Mineera shook her head in disbelief. Not even the psychic of the group could have guessed Banion's intent.

"Yes. The answer is simple. We have all the pieces to the puzzle of restoring order here in the Concrete Barrens and finding a nuclear weapon." Banion remained resolute. "The Slumlanders seek to destroy the rat tribe in order to obtain parts to fix their water purifier. The rats want to kill the Slumlanders to avoid losing their clean drinking water. If the problem of water purifier parts were to vanish, we could get our way."

"I'm not following," Jared said.

"Me neither."

"The Slumlanders sent several expeditions into the Shattered Isles, the chain of islands in the ocean to the west of here. The last expedition located the parts, but the expedition was destroyed before they could acquire them. We are going to steal a boat, find the purifier parts in the Shattered Isles, and end the war between the rat tribe and the Slumlanders. With the threat of being destroyed by the Slumlanders gone, I think Lavosi will still give us the information we need." Banion's plan was exasperating, but was, at least, a glimmer of reason in such dark times.

"That's impossible. The Slumlander leader has no idea where the expedition found the water purifier out in the ocean. There are hundreds of islands out there. What do you want us to do? Search them all?" Mineera retorted, finding the plan ridiculous.

"Would you rather go with the alternative? Kill the Slumlanders? Poison them to death?" Banion shot back.

He was right. Killing the innocent to spare the world from the Reaper Kai was the wrong thing to do. It didn't take much to convince the team to relinquish the idea of poisoning the Slumlander tribe. Though the task seemed impossible, it was a real solution as opposed to mass murder.

"What about Guillotine?" Tani asked, rubbing his bandaged arms, still feeling the sting of the cold blade slicing into his warm flesh.

"Screw him. I would rather risk getting ambushed by mercenaries, assassins, and Reaper Kai than to murder without recourse. If we lose our compassion, we will never make it to the end of this bloody road," the veteran warrior concluded. "We will have to risk Guillotine finding us once more."

The idea was rapidly growing on the rest of Nova 7. "I like this plan *much* better." Mineera smiled in excitement.

"Me too!" Tani concluded with a grin.

Extending his hand into their midst, Jared voiced his own exhilaration. "Here's to Nova 7!"

Each of them placed their hand upon Jared's.

"Here's to stealing a boat!" Tani giggled back.

As they laughed, the team's spirits lifted. They sat in the corner of the building, outlining their next course of action. As the chill of the night washed over the ruins, the fire that each of the companions felt igniting in their soul pushed away any hint of darkness. The task of locating water purifier parts in the vast expanses of the Shattered Isles was going to be nearly impossible, but it didn't matter. Nova 7 would spare their souls the torment of darkness   and   risk   their   lives   to   do   what   was   right.

Chapter 14
Agents of Darkness

A black swirl of energy rolled around him, a thick, cloying hurricane of shadow. The room was bathed in the fresh sunlight pouring into the tower through an open window. Father Vertigo, lord of the Reaper Kai, was near the window. Normally, the sun's rays would have bathed his pasty white body in glowing light. But in this instance, the sunlight was driven back by a massive surge of demonic power. The aged priest stood in the center of a maelstrom of chaos, pulsing blue, crackling light as shadows rotated around him, driving the very sunlight back.

As he extended his hands, the energy darkened, turning a morbid black color. The black crackling energy surged past his outstretched fingers, and the evil husk that was Father Vertigo basked in the demonic glory of all the twisted spirits held inside his body. As the wisps of darkness flooded him, Vertigo was jolted with a fresh wash of perverse power; the demons sang his praises as the energy filled the lifeless shell of his body.

The chilling blast of energy was receding with a wail. The haunting voices of the dead and the damned were still whispering inside the tower room as the demonic magic began to dissipate. Opening his eyes, Vertigo stared about with a piercing gaze; he was no longer alone. Bleating in submission, a Goat Minion was huddled near the doorway, having just witnessed the display of raw power.

Black rotten globes, the filthy dark spheres that were his eyes, watched the Goat Minion with loathing. "Why have you interrupted me?" Vertigo boomed out. The baying Goat Minion backed away, carefully sniffing the air with its pink snout.

Speaking the language of darkness, the Goat Minion bayed and bleated, quivering in the presence of the Lord of Terror.

Nodding in acknowledgement, the dark father spoke to the minion of darkness telepathically. Father Vertigo's powers were so strong, he could simply *will* the Goat Minions to do his bidding. Reading the mind of the servant, Vertigo discerned that a visitor had just arrived in the city of Rasheed, a visitor eager to have an audience with him. He sent a strong telepathic message to his servant, willing the beast to allow the visitor entrance into the tower. Bowing as he backed away, the demonic creature exited the room. Immediately, a form garbed in blood red robes entered the chamber and bowed before the lord of evil.

"Remove your hood," Vertigo ordered the newcomer, wishing to look upon her form.

As she pulled the hood from around her face, a shocking display confronted the Lord of Evil. Her hair, once a vibrant blond, was now washed with jet black strands. Her bright blue eyes, which once could bring any man to his knees, were now infiltrated by darkness, slowly changing to a misty gray color. The once glowing white skin of her face was now marred with wrinkles. Darkness had flooded the demonic maiden, and the skin below her eyes had sunk into the sockets, leaving black pits in its wake.

Her form had become grotesque as well. The skin and flesh on Marion's forearms had withered, tightening around the bones underneath like stretched leather. It was as if the dark queen had been withered by famine and starvation. Whereas once the pristine princess walked with perfect posture, she was now hunched over as if her spine had been bent and broken. The graceful gait that had distinguished her as possessing royal blood had become a labored stride, a pitiful walk which caused her to lurch forward with uneven movements.

Marion Toil was no longer a pristine vision of beauty. Instead, the once-gorgeous princess had transformed into something much more sinister: a dark queen consumed with hungry intent.

"I see you have found favor with our masters," Vertigo boomed, smiling upon Queen Toil.

Bowing in reverence, the dark queen smiled back. It had

been a long time since the tainted queen had been in the presence of her dark master.

Being obedient, Sister Toil dared not speak until spoken to.

"Ah, how wonderful, you have finally learned your place. It has been many months since we have been in each other's presence. Tell me, young Toil, how did our troops fare against the enemy?"

"The soldiers of Rust Spire were stalwart; they fought bravely and waged a desperate defense. The entire first wave of our troops was destroyed. However, the balance of power shifted as the sun set and the Abomination was able to take the field of battle. The enemies were driven back, and Rust Spire and all of the surrounding mines and petroleum reserves have been captured by Reaper Kai forces. The entire southern region of the Darken Realm is unstable, ripe for the picking. For all intents and purpose, *we* control the south." Smiling, the witch of darkness held a quiet, sinister grace about her. Nightshade had taught her much indeed over the past few months.

"And what of Nightshade?" Vertigo quizzed the queen of Rasheed, still absorbed by her pale, almost grotesque appearance.

"She is currently defending the area, subjugating all of the other surrounding kingdoms and towns. It will not take long before the entire southern reaches are enslaved."

"Excellent! And what of *my* direct orders? Did my courier reach your battle lines unscathed?" Vertigo was speaking with an almost supernatural roar. For one possessing such a frail frame, Father Vertigo was strangely powerful, and his voice was like a crack of thunder that pierced the very bones of those around him with an icy chill.

"Your courier was received. Soon after Rust Spire is reinforced and the surrounding region subjugated, Nightshade intends to head out into the wasteland to destroy the tribal village of Scarskin. The Abomination and one thousand Biogtech soldiers are preparing for the search and destruction of the village." Marion sneered, elated at the thought of the village of Scarskin being razed in searing flame and all of its inhabitants slaughtered without mercy.

"The Lost Tribe of Ceibla Moralis will not survive the cleansing fire. Our forefathers were unwise to allow these dissidents and traitors of our people to grow strong in the silence and passage of time. These rogue Reaper Kai will die. Any warrior who can withstand our psychic powers is a direct threat to our objective of total domination of the Darken Realm. Regarding the tales of the tribal Jared killing our priests and being unscathed by their power, the *only* explanation is that he is a direct descendant of our order. Only other psychics or peoples from our bloodline can be immune to psychic attack. Jared from Scarskin will perish, and his bloody defiance against our proud priests will come to an end. I will make that fool pay for his arrogance with the slaughter of those precious to him."

"Yes, my lord, their end is near." Marion bowed once more. Her reverence was refreshing to Vertigo. When last they met, she had been arrogant and foolish. Now she embraced the darkness, and was well on her way toward a glorious service in the name of evil.

"Any news of Nova 7?" Vertigo pressed.

"Yes, my lord, that is why I sought an audience with your greatness. One of my operatives, a mercenary named Guillotine, has made contact with Nova 7 in the Concrete Barrens."

"The Concrete Barrens?" The dark father spoke in a whisper. His mind was beginning to spin. The Concrete Barrens were on the distant, western fringe of the continent. It was disconcerting that this legendary team had pushed so far into the ruined wastelands of the Darken Realm. A dull twitch filled him, and a distant hint of fear clawed at his mind. "That is grave news indeed. I assumed they would have given up their foolish quest for nuclear retribution. No such weaponry still exists. Our agents and expeditions have never found a viable nuclear weapon or even a hint of such technology."

Falling silent for a brief moment, Vertigo pondered the situation, his eyes narrowing in suspicion. It made no sense for Nova 7 to be on the western fringe of the continent unless they had a convincing lead on a viable nuclear warhead. Coming to a resolution, Vertigo spoke once more. "If Nova 7 is that far in the wasteland, they must have credible evidence of their goal. Nova 7 must be close to uncovering a warhead." His voice rasped as he shook with anger.

"Guillotine is currently tracking them, my lord. He is alone and requires assistance. The radio transmission was sent from a town known as Green Isles, my lord."

"Send word to Rust Spire. Deploy ten squads of Biogtechs

and Reaper Kai priests, seasoned priests, to track Nova 7. The rest
of the tracker teams in the southern wasteland are to rally to the
Concrete Barrens immediately. I don't know how Nova 7 could
have eluded all of our spies and assassins, but they will not escape
us again! Have the first tracker teams that reach the Concrete
Barrens scour the area for any sign of their passing. If all goes well,
our Reaper Kai priests will be able to intercept the nuclear weapon
before they do… if one still exists!"

Vertigo was quivering with rage. The demonic spirits
trapped within his body began to howl as they sensed weakness
within him. Vying for control of his body, the demons within began
to battle amongst themselves. A scream erupted from him as
Vertigo's powerful will overrode them. Returning all the screeching
wraiths inside of him to silence, Father Vertigo gazed at Sister Toil
before him. "What great fortune would it be to locate a nuclear
weapon! Perhaps use it on the Iron Kai? Locating a warhead before
Nova 7 is your primary concern. If a nuclear weapon does still
exist, I want it."

"Your will be done, my lord. With ten fresh tracker teams
and eight already searching, Nova 7 will not elude us again."
Bowing once more, Marion prepared to leave Vertigo's tower rising
above the battered city of Rasheed. Turning away, she moved
toward the door. But Vertigo was not finished with her. The evil
priest *willed* the Goat Minion at the door to block her passage.
Obeying without question, the Goat Minion stationed itself in the
doorway, barring Marion's passage from the room.

Responding to the action, the dark queen of Rasheed calmly
turned once more to converse with Vertigo. "What is your will, my
lord?"

"Are your agents in Markov ready to strike?" Vertigo
quizzed the queen of evil.

"Yes, my lord, all of the assassins are ready to strike and kill
all the Mord Tech leaders. Say the word and their lives are forfeit.
My saboteurs and spies are also ready to cripple the Mord Tech
infrastructure." Having built an extensive network of spies and
mercenaries since the fall of Rasheed, Marion had agents of evil
infiltrating almost every remaining kingdom within the Darken
Realm, including the powerful Mord Tech empire in the northlands.

"Brother Feral is marching troops to lay waste to the Mord

Tech capital. Send word to your assassins; tell them to strike in three days time. With Feral only four days away from laying siege on Markov, the disorder caused by the death of the Mord Tech leadership will give us a sizeable advantage." Vertigo finalized the assassination of the Mord Tech leaders with a grim smile.

"Your will be done, my master. The Mord Tech empire will fall." Marion Toil, the great betrayer, bowed before her dark master and left the chamber.

The silence and solitude of the tower were refreshing to the dark lord. Brooding about, he was lost in thought, filled by a growing sense of empowerment. Vertigo was tightening the noose. With two of the greatest empires in ruins, he now shuttled his ambitions northwards. All that remained in his path were the foolish Iron Kai and the Mord Tech dissidents. With the destruction of the Mord Tech capital close at hand, the Reaper Kai war machine could focus on destroying the wretched Iron Kai, ancestors of old who had cast his people to shadow and oblivion in ages past. Vengeance for former disgraces was about to be unfurled.

The world was teetering on the edge of ruin, and the Reaper Kai were applying extra leverage to push it to the brink. As the legions of evil drew close to the Mord Tech capital city of Markov, Marion Toil sent encoded radio transmissions to her operatives stationed in the city. Darkness was about to be unleashed once more, and the carnage of war was moving into the northlands.

Chapter 15
Murderous Intent

The Mord Tech guards who responded to the kitchen maid's knock on the door eyed her with a look of familiarity. Opening the door, they allowed her entrance into the Grand Marshal's quarters. The guards were at ease in her presence and smiled at her with broad grins; one of them even winked in her direction. Lucinda had been in the service of the Grand Marshal for many months now, and had been preparing the mighty ruler's dinner for the same amount of time.

With a bubbly gait she walked, bringing a hot, steaming plate of food over to the Grand Marshal, the high ruler of the Mord Tech empire. He gazed at Lucinda with a smile. Her brown eyes shone back as she gingerly placed the food before him. Eyeing the food, the Marshal asked the kitchen maid, "What do we have tonight, my dear?"

"Leg of lamb with mint jelly and a side of herb roasted potatoes." Unwrapping the Marshal's utensils, she placed a knife and fork before the leader of the Mord Tech empire.

Standing at attention, the kitchen maid watched as the mighty Marshal started with a bite of the lamb, lightly covered in mint jelly. He smiled in appreciation, then set out to take a bite of the potatoes. Lucinda watched intently as the Marshal skewered a chunk of potatoes with his fork. Slowly, he raised the utensil towards his lips. Inch by inch it moved closer to his open mouth. Biting her lower lip, Lucinda focused upon the Marshal with utter concentration. His expression conveyed the exquisite flavor of the warm potatoes.   Smiling, he looked back at her once more. "Excellent meal, Lucinda. You are dismissed…"

The Marshal's words were cut short. With a deep gasp, his eyes grew wide. A gagging sound echoed within the room as he clutched at his throat. Lucinda looked on in calm fascination, moving forward to *aid* the mighty leader. As he coughed violently, the potatoes worked their dark magic, clogging his airway, blocking it tight. Gasping, he gagged several times, trying to dislodge the wedged food in his airway. Lucinda moved forward and slapped him several times on the back. The food left his airway but he continued to gag anyway. Something was horribly wrong; something was choking him, a foul toxin hidden in the food he had just swallowed.

With a frantic scream, Lucinda yelled, "Help him!"

The two guards inside the room rushed over and observed the Marshal in alarm. His face was already turning bright red with tinges of purple as he feebly clutched at his throat. His eyes bulging out of his sockets, the Grand Marshal was choking to death. The guards moved to help him, providing the opening Lucinda had been anticipating.

Not seeing the blade, both guards were caught totally unprepared. With a quick, deft move, Lucinda the kitchen maid jammed the honed steel into the throat of the first guard. The attack was so rapid and vicious, the other guard never even saw the violent deed. As the Grand Marshal felt dizziness take him, he watched in horror as his trusted kitchen maid butchered his guard. Blood erupted in a violent spray, and the guard clutched his throat as it flowed freely down his neck.

Pointing, the Marshal alerted his remaining guard a second too late. As he turned around, the guard was just in time to see the blood-soaked blade slam into his chest. Piercing his ribs, the blade hit its mark, tearing into his heart. With a defensive move, the mortally wounded guard went for his gun. Lucinda was prepared, moving in swiftly. The kitchen maid pulled his arm forward, extending it before him. She struck at the guard's elbow with her open palm, causing it to emit a sickening snap as she broke his arm backwards, snapping it like a twig. The gun fell harmlessly to the floor as the guard's shattered arm drooped limply. Taking advantage of the prone guard, Lucinda moved in behind him, placing him in a head lock. Utilizing all of her weight, she brought the tough soldier to his knees. He struggled but a few seconds, choking to death as his chest wound bled profusely. The second guard died quickly and Lucinda was on her feet mere seconds later.

Still choking and nearly purple in the face, the leader of the Mord Tech empire eyed the sinister assassin with wild fright. He knew his death was soon at hand. Lucinda moved forward, pausing only a few feet before the mighty leader. As she extended her hand, an unseen force squeezed the Marshal's throat. The psychic attack was so severe that the blood vessels in the Marshal's eyes exploded, blood dripping down his cheeks. With a grotesque crunch, the Mord Tech let out a final gasp as his windpipe collapsed under the sheer force of the sinister psychic energy. Collapsing forward, Grand Marshal Ignatius Dormain, lord of the Mord Tech empire, passed into death.

As her dress was now smeared in blood, Lucinda quickly disrobed and moved over to her serving cart. A clean dress had been stashed underneath, within a hidden compartment in the cart's base. She donned the fresh change of clothes, the sinister snake tattoos scribed on both of her forearms disappearing under the concealment of her long sleeved dress. Regaining her composure, Lucinda, assassin in the service of Queen Marion Toil, affixed a fresh smile on her face and pushed her serving cart out into the hallway.

The guards never spared her a second look as she exited the Marshal's quarters. She had been working in the courthouse and government headquarters for months, and to them, it was just another day.

"Hey there, Lucinda." One of the guards smiled with a flirtatious look as she passed by. "Get a beer with me later?"

"No, no, not tonight. I have *too* much to do. Maybe another time." Smiling back with a coy expression, Lucinda maintained her facade. Her calm demeanor and confident moves concealed her better than any disguise. Passing out of the main hall, the wicked kitchen maid walked into the servant's quarters. Without a second thought, she passed the other servants and left through the kitchen's back door, still pushing her cart.

The back of the government building opened onto an alley behind the enormous courthouse. Darkness had already taken the land as Lucinda rushed into the chilly night air. Spinning around, the Reaper Kai agent looked back at the courthouse one last time. With a wicked grin, she dismissed the building and what it stood for. There was no justice in the world, no true justice. In a harsh world, might made right; this was the true mantra of the Reaper Kai. Taking a final look at the government building, she shivered in the darkness and cold of the night.

With a resolute look on her face, Lucinda focused on the next part of her murderous mission. Opening the hidden compartment in her serving cart once more, the dangerous Reaper Kai assassin recovered a thick bundle of explosives. Clutching them tightly, she pressed on into the darkness of the Mord Tech capital city of Markov. The priestess knew she had to reach her next target quickly. If the body of the Marshal were discovered, it would compromise the second portion of her mission.

Avoiding the military patrols on the main avenues, Lucinda kept to the shadows and moved toward the southern portion of the Mord Tech capital city. Within a few minutes, she had reached her second objective.

A tall tower of receiving dishes and transmission equipment was her final target. It was heavily fortified, and for good reason. The communications complex was a critical nerve center for the Mord Tech empire, the means for coordinating the movement of the enormous war machine that was the Mord Tech assault force. Crippling the communications center was a crucial aspect of any invasion plan.

Crouching in the darkness, Lucinda eyed her target like a cat eyeing a scurrying mouse. The communications center was surrounded by tall, razor-wire fences. Four heavily armed guards walked around the building, slowly scanning the perimeter.

Her eyes moved back and forth, watching the guards hungrily. Pulling a pocket watch from her dress, she looked at it intently and tapped it with her finger, awaiting a signal to continue her acts of murder and sabotage.

A whistle blew in the distance, signaling the change of guard. The soldiers in the current Mord Tech patrol were leaving their posts at the communications center. As they marched away from the perimeter, the building was left unprotected for but a moment, but a moment was all the sinister Reaper Kai operative needed to reach her target. Rushing forward with a bundle of explosives in hand, Lucinda entered the perimeter of the fence line through a hole that she had cut the previous night. Crawling through the razor wire perimeter, the Reaper Kai breached the defenses and rushed to the safety of the building.

Another whistle blew. The change of the guard had ended. Four heavily armed Mord Tech guards, chemical tubing jutting from their necks, moved to secure the communications center. Seeing them advance, she sought to disappear and avoid confrontation.

Lucinda slipped inside the building, breathing softly. Tense from prowling around, she took a brief moment to collect her thoughts. Nodding her head with resolute clarity, she was once again prepared to finish her objective. She rushed forward softly, moving quickly past the heart of the communications center. A hallway ran the length of the building along the southern wall. It was lined with windows revealing all of the technicians within, sitting behind computer screens and intently monitoring all of the empire's communications. At the end of the hallway was a staircase leading up onto the second floor. With a bold move, the Reaper Kai saboteur rushed down the hallway. Not a single technician spotted Lucinda sprint up the staircase.

The top floor had been left totally unprotected. The foolish Mord Techs felt that *no one* could breach their defenses, and this was to be their doom. Fumbling around amongst all of the electrical equipment, the Reaper Kai operative located her target: an enormous platform housing a collection of electrical equipment and power supplies. She crouched and planted the explosives near the generator and transmitters. With only a two minute interval before the explosion, Lucinda knew that she had to escape quickly. As the timer on the bomb began to count back, she rose to her feet and moved toward the staircase leading back to the lower floor of the Mord Tech communications center. Reaching the staircase, the sinister saboteur stopped dead in her tracks, hearing a deafening boom from somewhere outside the building.

The explosion rumbled, lighting the night sky with a plume of rising fire. The power station near the river, in the southern portion of city, was rocked with violent blasts. Other wretched Reaper Kai operatives had just unleashed their own set of explosives. The power station erupted into a firestorm of debris under the destructive power of the demolition charges. With the main power station in flaming ruin, light and power went out immediately in the entire city, plunging the capital of Markov into violent darkness. It took a few seconds for emergency power to restore energy to critical defense systems, and a few buildings roared back to life.

A piercing wail ripped the air. Sirens used in times of war and natural disaster screeched. The city was under heavy alert. With civil defense sirens blaring, Markov was springing to life as all military personnel were roused from their barracks to answer the call. Cursing, Lucinda rushed down the stairs; if she did not hurry, escape would be impossible.

As she reached the bottom of the staircase, the bloody escapade of the sinister Reaper Kai operative ended. Only a few seconds after the blast at the power station, armed soldiers had flooded into the radio center. Catching sight of the woman in a dress charging down the stairs put the soldiers on instant alert. Training their weapons on Lucinda, the officer screamed at her. "Down now! Drop to your knees now!"

Fear filled her; it was time to fight or flee, and her route of escape had been cut off. With a jolt of panic, the sinister Reaper Kai began to channel demonic energy – it was time to fight. As she extended her hand, a lance of fire erupted from her fingertips with a fiery orange glow. The fiery charge hit the officer, exploding upon impact and incinerating his flesh. As the blast reduced their leader to a glowing cinder, the remaining soldiers opened fire.

Bullets whizzed past the Reaper Kai assassin as she fought back with all of her sinister might. An additional blast of hell fire detonated, killing another soldier in a fiery shower of smoke and demonic energy. The battle switched into high gear, and the Reaper Kai agent was heavily outnumbered.

Quickly, the soldiers managed to gain the upper hand. A crack shot Mord Tech infantryman took aim and blasted a burst of lead at Lucinda. The gunfire tore forward, striking her in the chest. The high caliber bullets tore into her flesh, throwing her back against the staircase. Having sustained terrible wounds, Lucinda, agent of darkness, ceased to stir. Though her life had ended, her mission had been a success. The demolition charges she had placed were still active, and the timer was counting down.

Securing the area, the soldiers pressed upwards onto the second floor of the communications center, leaping over the dead Reaper Kai priestess. Their advance was slow, a little too slow.

Another explosion tore the night; the explosives had detonated. A bright flare of fire boomed as a rising plume of smoke and charred metal was blown free from the top of the communications center. The communications room on the second floor was blasted to rubble, killing the advancing Mord Tech soldiers in a fiery blast.

The explosion was so powerful that the tall communications tower rising out the building began to shriek under the strain of the blast, its steel supports collapsing. The stress on the metal was too much to bear, the reinforced girders bent, and the entire tower collapsed, destroying the crucial hub of radio transmissions. All communications within the Mord Tech empire ceased immediately.

As the civil defense sirens screeched in the darkness, several more explosions lit the night sky over Markov. The Reaper Kai operations had been a complete success. The primary power, communications, and weapons facilities had all been reduced to rubble by other saboteurs. The Grand Marshal, head of the Mord Techs, had been slain, as had General McGivens, head of the city defenses. The city of Markov was dealt a powerful blow. Key portions of Mord Tech infrastructure, all crucial parts of a strong military defense, had been destroyed.

As the fires burned in the darkness of the chill air, a host of enemy soldiers was nearing the outer defenses of the city. Led by Brother Feral, the most sinister warrior leader of the Reaper Kai, a full army of ten thousand Biogtechs prepared to destroy the heart of the Mord Tech empire. As drums sounded in the darkness, the perimeter of the city came under enemy fire.

Chapter 16
Growing Shadow

The civil defense sirens were blaring. Darkness had overtaken the city of Markov. The main power supply had been destroyed by a series of well placed charges, reducing the main power facility to a smoking crater. Emergency systems had kicked in since the blast, and power was coming online in staggered intervals, lighting the city in an eerie glow.

The yellow emergency lights lined the avenues and byways, barely illuminating the civil defense forces scurrying about like a swarm of ants. As the sirens blared, the soldiers rushed here and there, seemingly in random directions. Most of the citizens of Markov were standing in the front of their houses, looking at the chaos as terror rose in their hearts. Was this finally it? Had the Reaper Kai finally set their sights on the Mord Tech empire?

Field Commander Deetric had been roused from his quarters only a few minutes ago. The Mord Tech guard detail which had arrived at his quarters was mysterious about their intent and rushed him through the darkness of the city with a frantic fury. As they pressed into the dark city, the blaring of the civil defense sirens was an ominous tone. Shuddering and cringing, he tried to force the horrid wail from his ears. With each screech from the sirens, his anxiety mounted.

Fires still flared about the city where key military targets had been destroyed, providing strong evidence that the chaos of the night was more than just a drill. Deetric knew conflict had marched into the northlands. Trembling, he thought about the horrifying prospect of war finally coming to the home front.

Light was emanating from the courthouse located near the center of town, next to the government buildings. Emergency power had been restored and the courthouse, the largest building in the entire city with power, was like a beacon in the night.

Charging toward the light, the armed detail of soldiers rushed Commander Deetric into the courthouse. Inside was total chaos. Mord Tech soldiers, augmented with drug implants, body armor, and automatic assault rifles, were battle-ready and running about. Around and around they rushed like a swarm of frenzied bees, protecting their hive as if a bear was rattling the precious honeycomb with its spiked claw. The level of activity within the building was extraordinary. Deetric knew without a doubt that it would be a night to remember, a day that would be forever etched in the history books.

As Deetric passed, all the soldiers eyed him with quick, darting glances, avoiding direct eye contact. The indecision in their eyes filled Deetric with a growing dread. Why were they regarding him so? As his heart suddenly began to lurch in his chest, his terror began to mount. Something big had just occurred, and now Field Commander Deetric was caught in the middle of it.

The soldiers rushed him toward the quarters of the Grand Marshal, the leader of the Mord Tech empire. An armed contingent of house guards nodded at Deetric as he approached. The door to the Grand Marshal's room was open and a swarm of soldiers were packed into the office. Breathing softly, Deetric entered the quarters of the Grand Marshal and prepared for the worst.

Blood pounded in his ears with a dull thud. A jittery feeling coursed through him and he was close to throwing up from the sheer panic filling him. Bright flashes of light filled his vision, causing him to feel dizzy. On the verge of a full blown panic attack, Deetric gripped his hands tightly and breathed slowly, trying desperately to slow his heart rate. Taking a minute to compose himself, he took a shaky step into the office.

Moving inside, Deetric saw the body of the Grand Marshal lying amongst the bloody remains of two bodyguards. He startled back, emitting a strangled sound that was almost a shriek. As his eyes grew wide with fright, Deetric felt his stomach lurch. The head of the Mord Tech empire had been assassinated.

"Commander Deetric." The Grand Marshal's head of security strode before him with a crisp salute. "We are in grave need of leadership. The Grand Marshal has been slain by an assassin."

Deetric was stunned by the scene. He could feel his cheeks start burning as they flushed bright red. Everyone in the room was staring at him in silence. Knowing this was no ordinary moment, he was gripped by indecision. In an attempt to formulate his thoughts, he stared around him, his head beginning to spin. "Where is General Morse?" Deetric quizzed. "Has anyone notified General Morse of what has happened here? He is second in command."

"General Morse was slain just before the explosions rocked the city," the officer responded in a dark tone. He monitored Deetric's response with growing anxiety. "He was assassinated in his home, shot through the head by his servant."

The weight of the world shifted. With the Grand Marshal and General Morse slain, the balance of power had radically changed in the Mord Tech empire. The next in the line of command to lead the entire empire was Field Commander Deetric.

"Commander Deetric, you are now the Grand Marshal, leader of the Mord Tech empire. What are your orders?" Standing at attention, the officer saluted Marshal Deetric. All the other soldiers followed suit and saluted as well, coming to attention before the newly crowned ruler.

As the blood drained from his stunned face, Deetric collapsed into a nearby chair. His brown eyes dropped to the floor and with a look of defeat. Deetric clutched his head with his hands, wanting to pull the hair from his scalp. Despite his desperate attempts to force the reality of the situation from his mind, Deetric was coming apart fast. His small form rested in the chair as his very will to live faded from him. The entire fate of a nation, a nation committed to a war, was now in Deetric's hands. Slumped like a withered man, the new Grand Marshal was gripped with indecision.

Seeing his lack of control, the officer pressed once more. "What are your orders sir?"

The civil defense sirens were still booming in the darkness. His mind swimming, Deetric reviewed the night's events. Saboteurs had destroyed key positions within the city and had assassinated the two most seasoned leaders of the Mord Tech empire. It was too much to handle. As he thought back to the fall of Rasheed and Rust Spire, eerie similarities began to mount. Just prior to attack, the mighty empires had been attacked from within or destroyed by plague. In was no coincidence that the Mord Tech leaders had been slain. Knowing that he had to act and act fast, the withered Deetric sprung to life, leaping from the chair with a bold gesture.

The soldiers eyed him with doubt. Deetric was a competent commander, but could he be a wise ruler?

"Pull everyone, civilians and military personnel, east of the river immediately. Tell the populace to take only weapons and provisions. Rig all of the bridges with explosives immediately," Deetric growled. "These acts of sabotage are just like the attacks made on Rasheed and Rust Spire. The Reaper Kai must be on the way."

His words were greeted with unanimous hesitation. One of the officers boldly stepped forward, standing at attention respectfully, but challenging his decision nevertheless. "The evacuation of the city is not necessary. We are not under attack. The Reaper Kai would not dare attack us. We are one of the strongest empires in the entire Darken Realm."

His temper flaring, Deetric moved forward to meet the officer's challenge. "We are under attack right now. You know what happened to Rasheed and Rust Spire? The enemy is at our doorstep. Evacuate everyone who is currently west of the river immediately, or be stripped of your command." Deetric's response to the defiant officer was forceful and commanding.

"Yes, sir!" the officer yelped, exiting the room to carry out the evacuation orders. Though the order was rash, Deetric was a staunch military commander who was not afraid to use his rank to intimidate his subordinates.

"I may not be able to rule an empire, but I know how to fight a war!" Deetric growled, surveying the officers in the room. The new Grand Marshal had just undergone a rapid transformation – from mouse to lion. Realizing that the enemy would never attack key infrastructure positions without proceeding with a full scale invasion, Deetric knew every second counted, and that the odds were pretty high that a Reaper Kai army was already nearing the edge of the city. "What of our communications?" he inquired.

"The saboteurs destroyed the entire communications center.

Nothing survived the explosions."

"What is the maximum range of any of our secondary transmitters?" the Grand Marshal asked.

"Thirty miles maximum, sir."

"Call for immediate aid from both Deshal and Carnak. March all of their troops immediately to aid in the defense of this city." One of the officers acknowledged the order and rushed from the room. "Is there any way to reach our reserve troops marching on Detro Tech city?"

"No, sir, they are too far beyond the borders of our empire."

"Send runners on foot with radio transmitters. Send enough to break through the enemy lines. Get our reserve troops back immediately. They should be no more than ten days forced march back." Pointing his finger at an officer, Grand Marshal Deetric finalized his next order.

"Yes, sir!" Another officer rushed out to meet the demands of the new Grand Marshal.

"Engineering teams are to get the city's main power supply back up immediately. After power is restored, get a suitable transmitter operational so we can contact the Iron Kai. Without their aid, we could be overrun." Deetric was feeling confidence fill him. "Go now!"

Yet another officer rushed to fulfill his orders.

"Send two thousand troops to the northern and southern districts. Get full artillery batteries ready. Station them along the river front. If any enemy targets are spotted, you have full authority to fire at will. I want full machine gun nests ready along the river front. Stagger them far enough apart so you can spray the entire waterfront with machine gun fire. You have thirty minutes or you will have hell to pay!"

"Yes sir!" Another officer smiled, empowered by Deetric's zeal, rushing out to make war.

"What about our air strip?" Deetric asked hopefully. Over the past hundred years, the Mord Techs had patched together three combat aircraft from parts harvested from abandoned military bases. The combat aircraft would be a sizeable advantaged in the coming conflict.

"Our assault aircraft have been sabotaged. The control panels and electrical equipment have been ruined. It will take months to repair them."

"Damn them!" Deetric boomed. The Reaper Kai plan to destroy the Mord Tech infrastructure was complete. A desperate blow had been dealt to the Mord Techs, and their ability to survive the coming conflict was deteriorating.

There was a brief moment of silence. In the lull, the sounds of the civil defense sirens blared in the night. Placing his hands behind his back, Grand Marshal Deetric moved over to the window and looked out. He could see Mord Tech soldiers rushing about like ants, preparing to defend the city from aggression. Watching their quick movements, Deetric felt relieved. His people were rapidly preparing to defend their homeland. Satisfied, he sighed soundlessly. As his tough eyes surveyed the city of Markov, a radio receiver crackled in the room.

"Lieutenant Clark!" the radio sputtered.

An officer responded immediately into the radio. "Go ahead."

"Soldiers on the western edge of the city just took down two Biogtech scouts."

"Repeat, officer, did you say Biogtech?" The lieutenant felt a chill run down his spine as he exchanged startled glances with the rest of his companions in the war room.

"Yes, sir, two Biogtech scouts have been destroyed on the western edge of the city."

Everyone in the room stared at Deetric with a look of astonishment. The new Grand Marshal had been correct. The assassinations and acts of sabotage had been cunningly planned to narrowly precede a Reaper Kai offensive. Feeling more confident about their leader but fearful about the sinister reality facing them, the military officers fell silent and held their breath.

Like a symphony, more radios sprung to life in the room. Other officers were being contacted. Over a dozen emergency radio transmissions were now bombarding the war room.

"Enemy troops sighted! We are taking heavy fire!" a frantic soldier's voice erupted from the radio. Automatic weapons fire roared from the instrument, and the sounds of screaming filled the room. The officer responded as other radios exploded with sounds of conflict.

"We are requesting artillery support at position 450.3.85.

Heavy enemy forces, hundreds of Biogtechs, are marching on Decator Street," another radio crackled.

Within a few seconds, all of the remaining officers were receiving radio transmissions from Mord Tech soldiers taking enemy fire.

The boom of artillery echoed in the distance. In the western fringe of Markov visible through the windows, fire leapt as Reaper Kai artillery slammed into the city. Orange flame curled into the air, radios crackled with combat reports, and the civil defense sirens wailed in the darkness. Gritting his teeth, Grand Marshal Deetric prepared for an all-out conflict. The siege of Markov had begun.

Chapter 17
The Siege of Markov

The first rays of the rising sun could barely be seen over the ruins of the once-mighty capital of the Mord Tech empire. Smoke, clinging heavily to the air, had settled over the battered city. The entire western fringe of the city had been burned to the ground. The Reaper Kai war machine had consumed much in its hunger. In its march toward the center of Markov, the enormous Biogtech army had ground to a halt at the riverfront. The cunning Mord Tech soldiers had blown up the bridges leading into the heart of the city. A colossal siege was now being waged as the Reaper Kai forces tried to break across the blood-soaked Ice Spurn river...

"Get down!" a soldier yelled. An inbound mortar shell arced across the river, whistling in the air as it roared toward the machine gun nest. Hurling themselves to the ground, the machine gun team prepared for the explosion. The mortar round struck the ground with a thud. A split second later, the explosive detonated.

A blinding explosion rocked the machine gun bunker. Burning debris and shrapnel were flung high into the air. A Mord Tech soldier who had been crouched only a few feet away from where the deadly charge detonated took the brunt of the blast. His shredded, lifeless body was projected into the river in a sickening spatter of gore. Other soldiers were thrown to the ground, having sustained varied injuries.

Blinking in the haze of confusion, a wounded Mord Tech soldier lay on his back, staring upwards at the sky. The soldier's head was pasty white and bald, and a series of surgically inserted clear tubes protruded from his neck and skull in several places. A sickly drift of smoke rose from the crater that was once the machine gun nest. Blinking several times, lacking the will to move, the soldier rested for a few seconds in the middle of heated, thick combat. As he stared at the smoke-filled sky, the screams of dying men around him were but a distant sound. The smell of earth, burnt and baked by the mortar, wafted through his nostrils.

Blinking several times, the soldier was suddenly aware that he was gravely injured. Pain flared in his chest. With dull fascination, the soldier sat up. Bullets zipped and whizzed past his head as he stared down at the wound in his chest. A fragment of metal, still warm from the explosion, was wedged in his flesh, just between two of his ribs. A trickle of blood soaked through his flak jacket; a dull red tinge of color dripped from the hole in his chest. With sharp, throbbing pain filling him, the soldier dreamily grasped the fragment of metal protruding from him. The fragment came free with a tug, as did a surge of blood. The deep wound released a jet of crimson fluid, which erupted in a gruesome spray.

As blood pulsed from the wound, the soldier stared around in shock. Unable to even think, the wounded infantryman blinked heavily, allowing the horror of combat to elude his senses as his body bled to death.

"Hit his trauma pack!" a scream rose somewhere behind him. There was a flurry of movement. Another Mord Tech soldier rushed forth and groped around hastily on the wounded man's back. As he fumbled with the device, a loud hiss filled the air. The medical pack on the wounded soldier's back had just been activated. With a whir, the infuser shot a fresh rush of greenish blue chemical into the tubes jutting from the soldier's neck and skull. Pumping quickly, the green liquid filled the tube and rushed into his body and brain.

The chemical cocktail burned like fire. As the potent, lifesaving brew pumped through his veins, the sounds of combat began to grow. He stared down as the greenish blue liquid erupted from his chest wound, mixing with the bright red blood. A second passed and the wound closed, instantly clotted by the powerful chemical agent. While the medicine healed his wounds, the cocktail of chemicals hit the soldier's brain with staggering results.

The haze of shock diminished in the blink of an eye. An aggressive blast of anger rocked the downed soldier. The chemicals were also stimulating his brain directly, filling him with clarity and focused rage. The wounded soldier eyed the enemy with a look of hatred. Fumbling for a weapon, any weapon, the soldier who had been on the verge of death only a few seconds prior was now healed by the noxious chemicals pumped into his body from the injection device strapped to his back. With a resolute, determined look, the soldier sought to slay the enemies of the Mord Tech empire. His weapon came to rest on his shoulder and the clarity of his condition allowed him to become a fine instrument of war.

Allowing his gun sights to come to rest on a Biogtech soldier across the riverfront, the soldier aimed and opened fire. A burst of quick explosions rang out as his bullets tore into the plastic neck plate of the robotic death machine. The mechanical soldier's wounds were marked by a spray of motor oil as it lost hydraulic pressure. Cackling but once more, the Biogtech let out a shudder as the machine of war ground on around it.

The weapons of the Reaper Kai army were formidable and the Mord Techs were severely outnumbered, but the advanced medical technology was saving hundreds and hundreds of lives. The Mord Tech army would have sustained double the causalities if it were not for the lifesaving trauma packs with which each soldier was equipped. Technology and advanced medicine were winning the war, giving the Mord Techs a sizable advantage over the poorly constructed Biogtech soldiers.

"Hold them back!" the Mord Tech sergeant screamed, pointing at the collapsed bridge. The enemy, strong in number, was marching forward, long steel planks in hand, with the intent of bridging the gap and allowing further attacks by the Reaper Kai forces to penetrate the defenses of the city. If the Reaper Kai forces rebuilt the bridge, there was nothing to stop them from overrunning the city with their superior numbers of troops.

Hastily throwing the smoking machine gun back on its tripod, the rattled Mord Tech soldiers regained control of the mighty weapon. With a resounding battle cry, the troops rallied as the machine gun roared to life, spraying a mighty storm of deadly rounds into the heart of the enemy troops. White hot tracer rounds leapt forth, allowing the gunners to aim their fire precisely. Sweeping back and forth, the gunfire pelted the enemy soldiers without recourse. Explosions of hydraulic fluid and showers of sparks illuminated the air as the mighty weapon mowed down the opposition.

Within mere seconds, the Reaper Kai advance to repair the bridge was torn asunder and stopped with militant force. Seeing the Biogtech troops shredded and unable to repair the shattered gap inspired the Mord Tech troops. Their cheers amidst the chaos were a moment of victory that would prove to be short lived.

A tremor rocked the battlefield. Even amidst the crash of explosions and the clatter of gunfire, everyone taking part in the conflict could feel the vibration in the ground. It was initially weak, but began to build like a surge of tiny earthquakes. More tiny tremors continued, providing evidence that something wicked was about to be unleashed. Taking uneasy breaths, the Mord Tech troops felt additional tremors as something unseen on the battlefield since ancient times moved through the back ranks of the Biogtech army.

A jolt of panic rolled through the front lines as the source of the tiny tremors moved forward. Marching in unison, emerging from the smoke, was a mighty, lumbering host of enormous Biogtechs. The heavy assault rigs, a terrible weapon of the ancients, came forward. Armored in thick black steel, the automatons were a fearful sight. Standing nearly fifteen feet tall, the Biogtech assault rigs boasted a mighty arsenal of weapons. Upon each shoulder of the robotic terrors was an enormous cannon. Servo motors covered in thick metal piping jutted from the assault rigs' back, powering the ammunition lifts to the mighty guns. The heads of the Biogtech assault rigs were covered in thick black armor, steel plating from which only a single eye peered out. Scanning the battlefield, the yellow robotic eye on each Biogtech assault rig scanned the conflict, as each robot prepared to lay waste to the defenders of Markov by targeting the entrenched soldiers.

"We have unknown heavies inbound!" the sergeant yelled into his radio.

"Repeat, sergeant! What are you engaged with?" the radio crackled back.

"Unknown Biogtech war machines! Send backup..." A loud thud rocked the landscape. The front line of the assault rigs opened fire. Upon each rig, the weapons erupted. The first gun hammered forward, firing a heavy cannon round at the entrenched soldiers, while the second gun on the rig's other shoulder chambered a fresh cannon round. It quickly boomed forward as the other cannon recoiled backwards, loading a fresh round into the breach. And so the guns fired, pounding back and forth like the engine on a locomotive, one gun firing as the other reloaded. In a slow, incessant rhythm, the Biogtech assault rigs laid waste to the defenders, their cannons booming and hurling deadly projectiles forward. The weapons were so powerful that they tore the defenders in two, leaving nothing but blood and gore in their passing. The colossal Biogtechs managed to push all of the defenders back in a doomed attempt to seek cover from their awesome power.

The tide of battle was quickly growing grim. The heavy bore weapons were firing mighty cannon rounds into the Mord Tech army with devastating results. Most soldiers hit with the cannon rounds did not survive, even with their medical gear strapped to their backs. As the heavy cannons dismembered and severed flesh, the enemy leadership ordered the shattered bridge to be repaired. A host of Biogtechs lumbered forward with steel planks and other materials to fashion a crude bridge.

While pinned down, the Mord Tech army could only stare on in horror as the enemy repaired the bridge which, in its absence, had kept them safe from further advance by the enemy. As the mindless Biogtechs worked intently, with a cackle here and there, the makeshift bridge came together swiftly.

"I need support now!" the sergeant yelled in horror into his radio. The bridge was repaired within minutes, and hundreds of Biogtechs slowly lumbered across. The first wave of enemy soldiers managed to secure the other side of the river quickly with the support of the heavy assault rigs. The Reaper Kai had breached the riverfront and were now making a considerable push into the Mord Tech defenses on the eastern side of the river.

As they watched the invasion of the city occur through the newly repaired bridge, the officers of the Mord Tech army began to chant battle songs. The chanting was drowned out at first. But the longer the war songs continued, the more soldiers joined in, rallying around their leaders with boisterous chants of their own. Fire lit their hearts as they cheered and chanted. A few moments passed, and the soldiers in the demoralized Mord Tech army had regained their courage.

The Mord Tech troops rallied to the sound of their battle songs. Abandoning their positions north and south of the newly repaired bridge, hundreds of tough soldiers pushed quickly toward the scene of combat. If the bridge was not retaken quickly, the enemy would overrun the rest of the city by virtue of their sheer numbers alone. The defenders knew the fall of their empire would occur that very day if the enemy gained a solid foothold. Already, several assault rigs and several hundred Biogtechs had positioned themselves in Mord Tech territory.

"Take them out!" A host of infantry rushed towards the scene of conflict. "Load and fire at will!" Crouching, a team of Mord Tech soldiers readied themselves to retake the east side of the riverfront breached by the enemy. A steel rocket launcher was mounted on a warrior's shoulder, braced to be fired into the fray. Two additional Mord Techs labored to load a deadly explosive projectile. Locking the rocket into the tube, the soldier slapped the shooter on his shoulder. Concealed from sight, the crosshairs of the weapon fell upon the chest of a Biogtech assault rig. The Mord Tech pulled the trigger with an involuntary grimace.

With a blast, the steel launcher rocked back as the rocket ripped forward, leaving a trail of burnt propellant. The assault rig turned just in time to catch sight of the missile streaking toward it. With its single eye tracking the inbound projectile, it tried to move out of the way, but it was too late. The impact was tremendous as the dense tip of the warhead breached the thick armor of the Biogtech. A split second later, the missile exploded. With a shudder, a white blast of flame and energy erupted. Small pieces of armor were torn free, as were the internal components of the mighty Biogtech. With a rain of shrapnel, the robotic terror lurched to the right and staggered. The mechanisms controlling the robot failed, the electrical control boards and internal supports torn by the explosion. Lurching sharply to the right, the Biogtech assault rig tipped and fell into the midst of the smaller horde of Biogtech troops. Several Biogtech soldiers were smashed into greasy spots of oil and shattered plastic as the monstrous assault rig crushed them.

Emitting a fierce war cry, Mord Tech infantry rushed forward, firing weapons in full auto mode. Hundreds of rounds thundered towards the mass of Biogtechs. Like a storm of bullets, the projectiles pierced the ranks of the Reaper Kai robots with staggering results. The attack was so severe that several dozen Biogtechs were felled by the fierce gunfire.

The staunch counterattack by the valiant Mord Techs was well placed. Within the course of a few minutes, the eastern side of the river was being retaken by the city defenders. The Mord Tech soldiers labored to push the enemy troops back, with the intent of destroying the freshly repaired bridge. Though the Mord Techs fought a bitter, heartfelt struggle, it would soon prove to be in vain.

Seeing the crucial struggle for the bridge, the Reaper Kai commander himself decided to take the field and crush the courageous defenders. The dark warrior known as Brother Feral, aggressive general of the Reaper Kai forces, screamed in anger, feeling a sadistic rage fill him with madness and bloodlust.

"With me!" he yelled in glee, holding his hands upwards as if to draw the might of the entire battle upon him. Hearing the call of Brother Feral, sinister agents of darkness sought to surround him. Brother Feral was a master of close combat and all in his army knew of his intent. The wicked Brother Feral was going to spearhead the assault himself, with demonic fury guiding his actions.

A crackle of foul, dirty yellow energy emerged from the far side of the bridge, the side held by the enemy. A demonic form strode forth. Dressed in blood-red robes, a beautiful vision of terror confidently pushed forward. The sadistic Reaper Kai war master, Brother Feral, strode onward, surrounded by the crackling, dirty yellow surge of demonic power. His robes open so all could survey his perfect chest rippling with seemingly sculpted muscle, Feral smiled as he advanced across the bridge and into harm's way. His face was young and filled with pride. His smile still firmly in place, he began to quiver in anticipation; reckless slaughter was about to be unleashed.

Gibbering like a pack of hyenas, a full complement of Goat Minions flanked Brother Feral, striding forward with him to battle. Their flashing red eyes and noses sniffing the air for the scent of blood, the sinister Goat Minions hungered, craving the foul misery of bloodshed.

In a bold move, Brother Feral pulled a long blade from a scabbard at his waist. The black steel scimitar was curved and notched like a saw. Smiling in glee, he walked into the midst of combat and began to butcher the defenders of Markov without mercy. The Goat Minions, smelling fresh blood, became enraged by the sweet aroma of death. Whirling as bloodlust took them, Brother Feral's personal troops waged a horrific war against the defenders of Markov, quickly turning the tide of battle in favor of the sinister Reaper Kai.

Chapter 18
On the Brink of Ruin

His eyes flashing with madness, the screaming Brother Feral strode forth, surrounded in an ominous field of yellow, crackling energy. His muscles bulged as his taut aggression was unleashed. With a bold move, the sinister Reaper Kai general rushed the battle lines of the Mord Tech army. His intent was to drive back the defenders of the city who were attempting to destroy the newly repaired bridge.

His movements were supernatural. In the blink of an eye, he was gone, seeming to blink out of existence and then reappear. One second he was forty feet away, charging the lines, and in the next, Brother Feral was behind enemy lines, in the midst of the defenders.

With an earsplitting screech of fury, the dark priest began to swing away with his saw toothed blade, hacking into soldiers under the guidance of an insane malice. A Mord Tech sergeant was ill prepared for the attack. Appearing right in front of him, Brother Feral smiled wickedly. The officer could barely comprehend what was happening before he was viciously attacked. Though he leapt back, the Mord Tech sergeant was too slow. Feral hacked into his neck in one fierce gesture. As the blood sprayed violently from his ruptured artery, the sergeant uttered a final death cry, clutching his throat. The Mord Tech soldiers witnessing the act were in shock. By the time their brains rationalized what had just happened, Brother Feral was *gone,* leaving them all stunned.

Spinning around, the defenders of Markov could hear other officers being slaughtered in their midst. Feral would appear for a brief second, kill a Mord Tech officer, then vanish. Chaos rolled through the ranks, as many were overcome by a paralyzing fear creeping over their minds. Morale was beginning to wane and panic was taking its toll upon the Mord Tech soldiers. With key officers dying all around them, it was only a matter of time before the Mord Tech army fled the conflict, exactly as Feral had intended.

As Brother Feral slaughtered the leaders of the mighty army, his complement of savage Najaszim, the hateful Goat Minions, spearheaded the assault, killing anything in their path at the base of the repaired bridge.

"Hold your ground!" a Mord Tech sergeant yelled. "Drive back those beasts!"

A horde of Goat Minions were on the attack. Taking advantage of the indecision caused by Feral, the beastly warriors moved forward with frightening speed. Their assault was so quick that the defenders of Markov were able to fell but a few with gunfire before their ranks were breached.

Like insane killing machines, the Goat Minions bayed and bleated as their red eyes flashed with hatred. Wielding an assortment of hand weapons, the Goat Minions set out to crush anything in their path. The front lines turned into a bloodbath. Feverishly killing with relentless zeal, the sinister beast warriors ravaged the troops. Blood and severed limbs began to litter the field of battle as they bashed their way through the core of the defenders.

All the while, Feral was appearing and disappearing inside the Mord Tech battle lines, targeting the officers. The plan was cunning, yet so simple; Brother Feral savagely slaughtered all of the Mord Tech officers as the Goat Minions demoralized the front ranks. Without leadership and with chaos reigning, panic was bound to buckle the defenders of Markov shortly.

Spattered in gore, the agents of darkness gibbered relentlessly, driven by madness. The savage Goat Minions were stained by the conflict and hungering for more carnage. Though they had sustained heavy wounds, the Goat Minions had no thought of self-preservation. Despite bleeding horribly, they continued to kill. A semi-circle of carnage formed around the bridge. The Reaper Kai army had achieved a strong foothold, allowing further reinforcements to breach the waterfront and use the repaired bridge to access the heated scene of conflict.

Meanwhile, Mord Tech infantry were firing across the river,

keeping the enormous black armored Biogtech assault rigs at bay. The constant bombardment from the heavy cannon rounds was tearing a wide swath through the Mord Tech army and hope was beginning to fade. The Biogtech assault rigs fired with a relentless rhythm, tearing through rank after rank of Mord Tech troops, and inflicting a staggering number of casualties. The rapid Reaper Kai onslaught was too much to handle. To make matters worse, the enemy had managed to push their artillery further into the city. Now within range of the waterfront, Reaper Kai artillery batteries opened fire.

A boom rocked the front line of Mord Tech soldiers. Enemy artillery slammed into the massed troops, causing heavy casualties. Flung high into the air, the smoldering remains of the slain rained down on the already frightened defenders, filling them with further panic. The tide of battle was turning rapidly in favor of the forces of darkness.

Grand Marshal Deetric listened to the sounds of carnage from his radio. With a grim expression, he shook his head in dismay. The enemy troops were too many in number, and with one of the bridges repaired, they were now swarming into Mord Tech territory. From his vantage point, the newly appointed Grand Marshal could see the smoke and flames of war destroying one of the mightiest empires in all the Darken Realm.

"One day..." he whispered, helplessly mesmerized by the war consuming the city of Markov. "Centuries it took to build this mighty empire, and the enemy has destroyed everything in one single day." The thought was staggering.

"My lord!" A wide eyed lieutenant came before the leader of the Mord Techs.

Deetric turned to face him with a grave look. "Report."

"The northern bridge is being repaired as we speak. The Reaper Kai forces are crushing the northern defenders with more of those enormous Biogtech assault rigs. What are we going to do?"

Deetric could feel the wolves nipping at his heels. The Mord Tech empire was finished. With a grimace, Deetric pounded his fists against the fine wooden desk, the desk behind which every

Grand Marshal had sat for the last four hundred years. The flames of war were desperate, calling for drastic action. Shaking his head, he knew that retreat was the only alternative: save what remained of his troops to fight another day. With wild frustration, Deetric pounded the desk several more times, wanting nothing more than to miraculously save his people with the force of his fury. But the energy he expended was in vain.

"How many of our boats are large enough to take out those bridges?" Deetric inquired.

"It matters not, my lord. The enemy is too strong. Even if we take out the newly repaired bridges, we cannot keep them from rebuilding. We need more troops!" the shaky lieutenant shot back in despair.

"We are not going to keep them from repairing the bridges. Send boats laden with explosives right below the bridges." The Grand Marshal looked at his military advisors in dismay. He was fighting an internal struggle, but had no choice. Desperate times called for desperate actions. With labored speech, Deetric gave an order with profound implications. "Destroy the bridges. We are fleeing Markov."

The order rolled through the room, leaving silence in its wake. The shocked military advisors looked utterly defeated. In the stillness of the room, the radios began to crackle with the sounds of conflict and the screams of dying soldiers. Fleeing the city was the only possible course of action. With stunned looks, the military advisors began to relay the orders; Markov was to be abandoned. It was a grim decision, but the only wise choice of action in such dire circumstances.

"My lord, will we ever return home? Will we ever be able to come home to Markov?" an officer questioned the Grand Marshall, his voice haunted.

It took a moment for Deetric to process the question, and his eyes remained distant, fixed upon the smoke rising from the riverfront. Finally, he turned to the officer. "I am not sure. At this point, I am not sure…"

The order to destroy the bridges went out immediately. A team of staunch Mord Tech engineers began to pack several barges and large boats with explosives. The plan was to send the barges under the bridges and detonate the cargo, thus destroying the bridges and all support pylons. As the order to abandon the city went out, the defenders of Markov were still waging a heated battle...

Smiling, Brother Feral appeared before the lieutenant brazenly rallying the troops defending the southern bridge. With a vicious slash, Feral assaulted the officer as he was in the midst of giving orders. The saw-toothed blade tore flesh, and the officer retreated a step as the weapon bit into his form. The attack was hasty, and the officer managed to recoil and avoid mortal injuries. With his stomach bleeding horribly, the officer cursed the Reaper Kai priest and charged him with a wild fury mounting in his eyes. His sudden action succeeded in stunning Feral.

"Kill him!" the officer yelled. Stunned Mord Tech soldiers viewed their valiant leader grappling with the sinister priest and charged to save him.

Unable to disappear while grappling with the Mord Tech lieutenant, Feral set out to slay the foolish officer immediately. He kicked his opponent in the gut, causing the Mord Tech officer to reel backwards. Screaming with rage, the vain priest blinked out, disappearing from the scene of combat just as Mord Tech soldiers came to the aid of their fallen leader. Though horribly wounded, the officer's expression was hopeful. Despite his obvious power, the Reaper Kai priest had a weakness, and therefore could be killed. Elated by his knowledge, but still fading fast, the Mord Tech lieutenant collapsed in shock.

As the officer crashed to the ground, the last thing he heard before blacking out was the sounds of conflict and battle around him.

The Goat Minions had finally been slain, but the cost of this action had been high. Over one hundred soldiers had been slaughtered by two dozen Goat Minions. Though the cost in human life was grave, it paled in comparison to the loss of morale within the Mord Tech army. Hundreds of soldiers witnessing the carnage brought to their countrymen had fled the scene of battle, wanting nothing more to do with the conflict of war. The defenders around the southern bridge had buckled, and dozens of Biogtechs were crossing the bridge every minute, bolstering the Reaper Kai presence in Mord Tech controlled territory.

The situation was dire. The number of Biogtechs now in Mord Tech territory was nearing one thousand. The heavy assault rigs were blasting wide swaths through the Mord Tech troops, with scattered cannon rounds killing several soldiers every second. The defenders holding the southern bridge needed a miracle, and they needed it fast, or they would be overrun in a matter of minutes.

Lumbering up the channel came hope. Several barges covered with Mord Tech troops pushed up the river, toward the southern bridge. Sounding fierce battle cries, the soldiers upon the deck of the boats opened fire on the Biogtech soldiers crossing the bridge.

Watching the advance of the ships, Brother Feral, having fled combat, eyed them with scorn. His psychic abilities extended to read the intent of the Mord Techs. Knowing that their objective was to destroy the bridge, Feral ordered his troops to stop the boats at all costs.

Ignoring the Mord Tech defenders, all Biogtech troops within range of the bridge turned their attention to destroying the barges. Thousands of bullets were now directed at the boats lumbering up the channel. The Mord Tech defenders on deck responded with more battle cries as a swarm of bullets hit the ship. It took but a few brief seconds for the torrent of gunfire to fell all of the soldiers on the deck; most were torn by dozens of bullets, leaving only bloody traces of their former selves behind.

The cabin of each ship then became the focus of the Biogtech barrage. Cannon rounds tore through the steel hull plates, scattering the crew as they ducked for cover. Artillery rounds, hastily targeted, came screaming down into the river with splashes and explosions. The Reaper Kai artillery teams had responded to Feral's orders, trying desperately to sink the ships before they reached the bridge.

The barge leading the assault lurched strongly to the right. The crew had been slain and the boat was out of control. Crashing into the walls of the canal, the Mord Tech ship ground to a halt. A split second later, an artillery round slammed into the deck of the barge, shredding the deck plate with a deafening boom.

The other two barges lumbered forward undaunted. The crew on the ships knew that their lives were forfeit, but it didn't matter; their sacrifice would ensure the survival of all that remained of the Mord Tech army and its civilians. With grim determination, the crew of the other two barges persisted in their mission. Focusing on the second ship in the convoy, enemy gunfire tore into the vessel, resulting in frightening damage. Cannon rounds, gunfire, and artillery slammed into the ship, but it persisted in inching onward. With only a few crew members surviving, the barge neared the bridge, its deadly cargo of explosives lying in deadly wait.

The push to destroy the bridge was in vain. Two artillery shells struck the ship, piercing the deck plate, and both exploded in the hull of the barge. The explosions were so severe that the hull buckled and split as fragments of steel were blown clear of the barge. Taking on water, the bow of the ship tipped forward and then ceased to move.

Two of three ships had been brought to a halt, but one more still remained, and witnessing the demise of their countrymen had only made the crew all the more stalwart and resolute.

Swerving around the two boats blocking the river, the crew members expertly avoided the wreckage. The Reaper Kai firepower directed toward the final boat was formidable. Many artillery shells slammed into the river, hundreds upon hundreds of bullets were directed at the ship, and the Biogtech assault rigs shredded the barge with heavy cannon rounds.

Despite the onslaught of aggression, the crew managed to escape ruin. Lumbering forward, the steel-clad barge moved towards its target. Feral cursed the barge as he retreated, knowing full well what was about to happen.

Screaming in defiance, the bridge crew prepared to meet their maker. The barge reached the bridge and passed beneath it. As it did, the Biogtechs on the bridge fired in full auto mode, their weapons humming away in an attempt to stop the ship. The Mord Tech soldier on the ship's bridge smiled, clutching a detonator in his hand. With a grim look, he closed his eyes and pressed the trigger.

The explosive laden barge was ripped by a terrifying explosion. Several hundred pounds of explosives detonated in the cargo hold. A bright white surge of energy rushed up. The blast was tremendous, and a mighty boom rocked the landscape. Parts of the barge vaporized, utterly obliterated by the blast. The energy unleashed erupted upwards, consuming the southern bridge in a rising plume of destruction. The support beams buckled and curled, melting completely at the center.   The bright orange inferno mangled the bridge, reducing it to a heap of melted slag and debris. Everything on the bridge – concrete, metal, Biogtechs and all – was destroyed and so hopelessly annihilated that nothing remained but steaming vapor.

As the boom rocked the battlefield, the Mord Tech engineers reached the northern bridge as well, detonating another barge filled with explosives, and turning this bridge into burning slag and rubble as well.   With both bridges destroyed, Grand Marshal Deetric looked out from the government building and let out a sigh of relief. The Mord Techs had managed to cut off the route into the city for the time being, giving Markov crucial time to evacuate.

Speaking into the radio, Grand Marshal Deetric gave the order for a full retreat. The Mord Tech empire, one of the strongest in all of the Darken Realm, an empire that had grown strong for hundreds of years, had now been destroyed. With none but the Iron Kai remaining, the Reaper Kai's goal of absolute dominion was now a feasible reality.

Chapter 19
A Heavy Burden

The often aggressive waves of the mighty lake were calm. Emperor Gunther's arms were folded behind his back, and he was watching the lake with a momentary sense of serenity. It had been weeks since the Emperor had taken a break from the war against the dread Reaper Kai. Gunther had made a promise to himself not to listen to any of the radio transmissions coming out of Detro Tech City. It was a peaceful day in which Gunther could reflect on his own life without the carnage of war and the heavy burden of leadership.

Moving to his throne, Gunther recovered his mighty beer stein. Clutching it lovingly, he smelled the strong brew held within. Smiling, the mighty leader tipped back the stein. The malted beer greeted his lips with hungry anticipation and slid down his throat, warming both his body and soul. As the liquor worked its magic, filling him a lingering, drowsy calm, Gunther sighed and smiled. Allowing the moment of serenity to fill him, he sat alone, atop his mighty throne, watching the lake below from the windows of the mighty Truce Hall. For that single moment, he was truly alone and it felt wonderful. No advisors were stating their cause, no military leaders were frantically trying to glean information from him, and no military reports full of casualties or other horrendous statistics were being unceremoniously tossed into his lap.

Taking in the peaceful calm of the morning, Gunther smiled and simply existed, without worry or distress. Letting his mind wander, the mighty monarch was nearly dizzy with the elation of momentarily abandoning his responsibility. But his calm day was to be short lived.

A burst of movement met his eyes. A military advisor rushed into the Truce Hall, his face entrenched with worry. With panic in his eyes, something to which Gunther was becoming more accustomed in his advisors, the man almost tripped halfway across the Truce Hall. The uneasy gait and terror on the advisor's face put Gunther on edge. The information the advisor was carrying was not going to be pleasant. With tension filling him, the slumped Gunther, in a slight drunken stupor, straightened in his throne. Sighing, with a grave look coming over him, he braced for another round of bad news.

"This is my day off," he grumbled at the advisor.

"My Lord..." the advisor stuttered, unable to speak properly. Gunther felt a knot form at the back of his head. Even at the worst of times, the advisor was more presentable than he was now.

"Calm yourself," Gunther said firmly. Another flash of movement brought the monarch's eyes to the entrance of the Truce Hall. All of his other advisors and military contacts were crowded around the entrance, but dared not enter. As Gunther eyed them, a sharp apprehension filled him. All the people standing in the doorway had panic-filled expressions. Collecting his thoughts, Gunther knew the report was going to be the worst of his days as Emperor. "What is your report?"

Shaking as his eyes darted back and forth, the advisor held a parchment in his unsteady hands. Unable to comprehend what was written upon the parchment, he began to stutter once more.

"The Mord Techs, my Lord..." Breaking down, the advisor could not continue. The fear on his face was so dire that Gunther was becoming drastically concerned.

"What has happened?" The tense knot at the back of his skull was growing and his stomach lurched uncontrollably.

"The Mord Tech empire has been destroyed."

Gunther blinked several times after hearing the dire news. It took a few moments for the horror to traverse his overburdened mind. "Say again?"

"The Mord Tech empire has been crushed. An army of ten thousand Biogtechs led by Brother Feral has just taken control of Markov. Another Biogtech army, five thousand in number, is currently burning and enslaving the other eight Mord Tech townships. All that remains of the Mord Tech empire is fleeing and marching to Stonen, my Lord."

"That is not possible!" Gunther yelled in defiance, standing up from his throne, his red beard quivering with rage. "We have contained the enemy capital for months. It is not possible that fifteen thousand Biogtech soldiers breached our battle lines!"

"I cannot explain it, my Lord. It is just like the fall of Rust Spire. The Reaper Kai must have had sizeable reserves stationed all over the Darken Realm prior to the beginning of the war."

"I cannot believe that. We have been made fools. The enemy is much more cunning than I ever imagined. Get in here now!" Gunther roared at all the advisors crowded around the doorway to the Truce Hall.

Hurriedly moving inside, Gunther's support team was not enthusiastic about the new task at hand.

In the silence of the chamber, all in the Truce Hall could hear their own hearts beat. The eerie quiet gripping the room went hand in hand with the gravity of the situation. All were dreading the revelation of yet-untold facts. As Gunther took several minutes to collect his thoughts, no one dared speak or move.

Focused by the eerie silence, Gunther's thoughts became resolute; he knew what needed to be done.

"A blood feud that has been raging for more than three hundred years is about to come to an end. Our ancestors gave rise to the wretched Reaper Kai, and now the entire Darken Realm is on the verge of ruin. The mistakes of our ancestors are being felt by all." Taking a brief second to collect his thoughts, Gunther shook his head in dismay. "We are truly alone. All the other great houses of the Darken Realm have been annihilated. The remnants of Rasheed, the Steel Crag Mining Guild, and now the Mord Techs have all pledged their support to our banner. We are the final empire that stands in the way of the Reaper Kai and the total destruction of all free peoples of the Darken Realm. Our resolve and determination to survive and destroy the enemy has never been stronger. We no longer only have a responsibility to ourselves to see the destruction of the dread Reaper Kai, but also a responsibility to all the other races and empires that now cease to exist or are on the verge of ruin. We are the final force to hold the enemy back and bring justice to all the misdeeds of our fractured race. We, as the Iron Kai, now need to slaughter and destroy the mistakes of our forefathers." Everyone in the Truce Hall listened intently to their monarch. With intense fire filling his eyes, Gunther's voice rose in tone and boomed forth, turning all eyes to him by the sheer force of his words.

"Do not fear the coming darkness. With justice and virtue on our side, evil cannot stand to triumph. In these dark days, we must all hold together and embrace the strength of our convictions and the courage of our hearts. Do not dismay! Do not feel terror! We have prepared for centuries for just such a conflict. Our army is the finest in the land, a well oiled machine, an instrument of raw power. In the end, the Iron Kai will lead all the free peoples of the Darken Realm to victory!"

The speech was inspiring and the military advisors and officials were set at ease. The stakes had never been greater, but Gunther's staunch defiance of the Reaper Kai advance filled all in the room with heightened resolve. Each member of Gunther's military team was becoming confident, and the fear of utter ruin was lifting from their minds.

"What is our next course of action, Lord Gunther?" an advisor questioned the mighty monarch.

"The soldiers pouring in from the Mord Tech empire are to be used to reinforce our siege of Detro Tech City. With their increased support, containing the enemy and keeping our advance should not be difficult," Gunther replied confidently.

"What of the other recruits, my Lord, from such places as the Steel Crag Guild and smaller provinces coming under fire?"

"Use them to reinforce this city, our capital. If we lose our ability to produce weapons and supplies, our chances of survival will be slim." After giving his orders, the emperor moved on to other topics. "What news from our commando teams headed into the Gold Road? Have they destroyed the tunnels under the mountain with explosives? Have they effectively cut off the supply route from the mines and petroleum reserves of the former Steel Crag Mining guild?"

"Not yet, my lord. The commando teams, led by Globulus of Rasheed, are currently moving to strike and sever the supply route from the Steel Crags. If all goes well, the Reaper Kai will be unable to move shipments of steel and petroleum out of the mountains. It will take them many months to clear the ruin the commando teams are moving to unleash."

"By severing their supply lines and cutting off their resources, we should make it nearly impossible for the Reaper Kai to continue their production of troops."

"Agreed, my lord. Severing the enemy supply lines should be a primary concern. Now that the matter of the commando teams has been addressed, what are your thoughts on the large number of Biogtech soldiers?" the military advisor quizzed Gunther, and everyone in the Truce Hall moved a step forward to hear their emperor's response.

"The number of troops that sacked Rust Spire and Markov mystifies me. I cannot fathom how the Reaper Kai managed to get thousands upon thousands of troops beyond our battle lines to far fields of combat without us knowing. We must double our efforts to ensure no more enemy troops escape," Gunther concluded, to the agreement of all in the room. Without credible evidence of how the enemy was marching thousands of unseen troops through the battle lines of the Iron Kai, the danger of surprise attack was an enormous threat.

"Agreed, my Lord. We will double our efforts to ensure no other forces leave the Reaper Kai capital. We will set additional scouting vessels on the lake. Perhaps the Reaper Kai are moving the soldiers by boat. We can scarcely afford any more surprise attacks."

"What of the rumors of new Biogtech troops on the field of battle?" another military advisor strode forward and quizzed the mighty monarch.

"You have heard all that I have," Gunther responded calmly, hoping that the parchment sent from the Mord Tech empire would yield more information.

"The parchment speaks of a new model of Biogtech that took the battlefield of Markov. Something never seen in combat thus far, even in the Reaper Kai capital. The new model of Biogtechs was heavily armored and carried a pair of shoulder-mounted cannons. According to this communication, the Biogtech assault rigs were extremely formidable and nearly impossible to destroy." The war advisor was now clearly calmer as his eyes scanned the parchment.

"We will move additional heavy armament to the Reaper Kai capital. Thus far, we have not required such weaponry since the normal Biogtechs are not difficult to destroy. The only thing that puzzles me is why we haven't seen any of these assault rigs, especially in Detro Tech City? It just doesn't make sense that the Reaper Kai would allocate their best troops somewhere other than their capital which is currently under siege." The mighty emperor shook his head in incomprehension, stroking his mighty red beard.

"True, my Lord. I am also concerned with this topic. It also makes me wonder what else the Reaper Kai have up their sleeves. The large number of enemy soldiers is also frightening," the field commander responsible for the defense of Stonen added with worry in his eyes.

"The number of enemy troops engaged within the Reaper Kai capital, and the ones holding Rasheed, Rust Spire, and now Markov have our own troops severely outnumbered. According to our new estimates, there are at least thirty thousand Biogtechs currently on the field of battle. Any more surprise attacks will add to this already horrendous number. An attack on Stonen would be nearly impossible to repel. A miracle is needed to ensure our victory. At this point we cannot afford Nova 7's failure. The success of Banion O'Neil and his team is essential. Is there any additional news about Nova 7?"

"No, my Lord. Their location is currently a mystery. I must impress upon you that our own Iron Kai scouting teams have scoured the ruins for countless generations and no viable nuclear weapon has ever been located. The success of Nova 7 is a long shot."

"Long shot? They have survived countless battles where even our most battle-hardened troops would have failed. They will succeed. I have no doubt. The entire fate of the Darken Realm is now hinged upon their success." Gunther was staunch in his defense of Nova 7, but none of his advisors shared his foolish optimism. "Without the destruction of Detro Tech City, we will be overrun; it's just a matter of time. We must trust in their success."

All in the Truce Hall looked at their emperor as one would view a foolish child. Most felt that his brazen hope for Nova 7 was dangerously misplaced. Finding a nuclear weapon was not a viable option. But even as Gunther's advisors secretly scoffed Nova 7, the mighty emperor maintained his faith. The lord of the Iron Kai was convinced that the bold mercenary team was hot on the trail of a nuclear weapon. In his mind, the impossible had already been made possible by the legendary team. It was just a matter of time before the lauded heroes would return from the wasteland with the holy grail of ancient artifacts. However, this precious time was quickly running out with each passing day.

Chapter 20
Under the Threat of Ambush

The rusted railing was warm to the touch. Mineera supported her weight upon it, bending forward to look over the edge. Down below, bobbing up and down on the ocean waves, was a collection of boats moored to a crudely constructed dock. A number of sailing folk were going about their business, stowing supplies or offloading wares to be sold at the ocean outpost. Uninterested in their daily duties, Mineera allowed her attention to drift elsewhere.

She sighed and threw her head back, letting the warm rays of the sun heat her skin. Closing her eyes, Mineera smelled the fresh salt breeze and felt the wind push through her dark hair. As the interplay between the cool wind and warm sun tickled her face, the day felt truly peaceful. Smiling, she opened her eyes and looked at the world around her.

Mineera was startled to find a large white seagull on the railing, only a few feet away. Bleating suddenly, the bird began to shriek at her. She grinned, understanding the ocean bird was trying to beg a scrap of food from her.

"I don't have anything for you," she said in a soft tone.

The bird looked back at her, cocking its head from side to side, peering at her from the corner of its eye with its head crooked. It shrieked again, more insistently. Mineera smiled in amusement. The bird could sense her detachment, and knew all too well that a meal or scrap of food was extremely unlikely. With a nasty squeal, the gull flapped its wings at Mineera in defiance, continually bobbing its head up and down in an attempt to intimidate her. She shook her head, almost wanting to laugh at the silly creature.

Finally, it flew away, leaving her alone upon the catwalk.

Amused by the day's small adventure, her attention shifted once more. It had been a while since she had seen the rest of her companions.

Spinning around, Mineera looked for the rest of Nova 7. Scanning the ramshackle buildings and corroded catwalks, she tried desperately to locate the rest of her team. Try as she may, none were in her line of sight. While scanning the buildings and patrons of the ocean refuge, Mineera caught sight of an unusual woman, crouched on a catwalk with a fishing rod. As her eyes passed over the fishing woman, a clenched sensation washed over her. The feeling was so surprising that Mineera was set on edge. As she observed the strange woman carefully, a wave of nostalgia came over her. The woman holding the fishing rod was familiar, a little too familiar. Fragmented images came into Mineera's mind. She experienced a wash of fear and panic. The woman was gripped by evil and seemed familiar. Blinking several times, the psychic was seized by an eerie vision. The fishing woman was dressed in a long sleeved shirt. The sleeves were loose, and Mineera caught sight of a black ink tattoo upon her wrist.

A tremor filled her and a scream began to rise in her throat. The tattoo was that of an ominous serpent, branded on the woman's flesh. The feeling of nostalgia and sickening emotion became clear. In a frantic attempt to warn her companions, Mineera yelled in panic, "Reaper Kai…"

Shaking violently and sensing someone huddled over her, Mineera opened her eyes, escaping the dream. It was dark, still nighttime, and a form was hunched over her. The hands of the person were on her shoulders, gently rocking her back and forth. Panic filled her and she flailed her arms about, trying to drive off the person touching her in the blackness of the night.

As her eyesight cleared, she stared at a hideous creature. Its beak was hooked, and a fiery red glow shone in the center of its black eyes. The strange form was charred and small, mostly gray in color, with tinges of black. Blinking several times, Mineera struggled to gain a grip on reality. The creature before her was the strange bird totem, hanging down from Jared's neck. The tribal warrior was crouched near the psychic maiden with his hands lightly pressed against her shoulders. Heaving a sigh of relief, Mineera sat up and stared at Jared.

"You were yelling in your sleep. I didn't want you to wake the others," Jared said earnestly. He seemed truly concerned and Mineera was set at ease.

"I was having a bad dream..." she said, placing a hand upon her forehead. Shaking off the stifling feeling of the frightening dream, Mineera felt her heart rate begin to decrease as the panic abated.

"Yeah, I thought so. It's still a few hours before dawn. Go ahead and get some sleep, it's still my watch." The tribal stood up to move away once more. Though he had been cordial with her over the past few weeks, he had still not forgotten her treachery. Wanting to escape her presence, he tried to slip away into the darkness.

As the young warrior moved away, Mineera grabbed his leg, urging him to turn around. "Sit down for a second," she coaxed the youth from Scarskin.

Jared eyed her with suspicion. His eyes glanced at Tani and Banion sleeping at the far side of the camp. With obvious reluctance, Jared finally sat down near Mineera and stared at her with a strange look on his face. Ever since her total breakdown in the eerie tunnels of the Iron Gate ruins, Jared had given her a wide berth, not necessarily being rude, but none too comfortable with her either. He would only converse with her if he deemed it absolutely necessary. This interaction was already breaking the tribal's rules.

"I wanted to say that I'm sorry. I'm sorry for what I did to you back there." She was having a hard time looking him in the eye. Finally, she gained enough courage to gaze at him directly. The tribal youth looked back uncomfortably and then averted his gaze.

"I was in a bad place, a place where I will never go again." She spoke in an honest tone, trying to sound sympathetic and truly apologetic. "I know how you must *feel*, being betrayed by a companion." Mineera was not being completely frank with the tribal youth. The strange bird totem made it impossible for her to truly sense his emotions. While he wore the totem, her ability to read his emotions was effectively blocked. Instead, she was using her experience of human nature to try and understand him.

"How I feel? How could you possibly know how I feel?" Quickly becoming agitated, Jared spoke with venom.

He was noticeably upset, almost to the point of losing control. It didn't take psychic powers to figure out he had more on his mind than just Mineera. Narrowing her eyes, she sought to pry additional information from him. Slowly and carefully, she watched him, not wanting to press forward too quickly. After a moment of silence, Mineera addressed him softly.

"I pray every night not to dream. I pray to heaven itself to spare me the suffering of my visions. No matter how hard I try, the visions come to me. I am tortured by the dream world. Though I cannot know exactly how you feel, Jared, I know what it is like to be tormented, more so than you could ever imagine."

Feeling a bit taken aback by her openness, Jared looked at her with unguarded eyes. The longer he gazed into her eyes, the weaker he became. Something terrible was consuming the tribal, something he had been hiding. Shaking his head, fighting back the tears, he tried to conceal his vulnerability.

Seeing his distress, Mineera reached forward compassionately and touched his arm. Feeling comforted, he blurted out an ominous statement. "Odd that you should mention being tormented by your dreams."

Blinking several times, she felt a lump rise in her throat. Pushing onward, Mineera sought to discover the dark secret. "What is it? Have you been having nightmares?"

"You could say that." Shaking his head, Jared was rattled. Looking over toward Banion, the tribal made sure that Banion was asleep. When he turned back to Mineera, the floodgates opened and the young warrior began to tell a chilling tale.

"I thought it was just coincidence. You would ramble on and on about how I was the enemy and would bring about the death of the rest of Nova 7. I thought you were crazy, just having bad dreams." Shaking his head, he fell silent for a brief moment. "I was wrong, horribly wrong."

Stunned, Mineera gripped his arm tightly. "What are you talking about?"

"I have had the same dream over and over again. I thought it was mere coincidence at first, but now... I am not so sure," Jared confided in Mineera.

"Tell me what it is. Tell me what you are seeing," she coaxed as an eerie fright washed over her.

Closing his eyes, the tribal spoke in a near-whisper. "I'm not sure where I am. The walls are strange, as if they are made of metal, corroded metal. My body is being wracked by anger. Something is filling me with rage, and I have no idea what is going on. It feels like reality has collapsed. With frantic terror, I look at my hand, and I have my sword. The blade is stained, horribly smeared in gore. There's black blood on it, and a foul smelling stench rising from it. In the dream, all of you are gone, all of you are dead. Banion, Tani, and you are all dead in the nightmare. I am standing among your bodies, completely alone, filled with an anger that I cannot control. After seeing this dream over and over, I'm beginning to think that maybe I am the enemy. Maybe I am beginning to lose my sanity."

As he finished his story, Mineera was taut with anxiety. The vision was strangely similar to her own, the one that she had seen dozens of times.

"I am worried, Mineera. I feel strange, almost as if I'm changing somehow. I'm worried that what happened to you is happening to me. What if I am beginning to lose my mind like you did? What if the evil spirits that sought to ruin you are now infesting me?"

Stunned by his confession, Mineera was taken aback. The proud tribal always appeared in control. She was astonished to find out that this was far from the case. Jared shared her own fears. While previously, she had felt separated and distant from him, a strange sense of kinship now began to form between them. Nodding as if confirming that she understood him, Mineera tried to console him.

"Dreams are scary, terrifying for me. I thought that no one else could ever understand how I felt. I dream the same horrible things night after night and I am powerless to stop them. Most of them are portents of the future, but many are not. Many are just dreams, strange images that I cannot control."

"How do you deal with it? I see you night after night, wracked with terror, talking in your sleep. How do you deal with dreams?" Jared quizzed her, feeling more confident about his decision to confide in her.

For the first time in her life, Mineera was not seen as insane, as helplessly driven by dark dreams, but rather as a mentor, someone whose rough experiences could be used to aid others. It was a strange dichotomy, moving from the learner to the teacher in a matter of seconds, from someone lost to one who was now a guide. Coming to her senses, Mineera used her own experiences to comfort the rattled warrior.

"It's hard to describe. When you have nightmares night after night, it taxes your mind. It seems that nothing can make the world better, and you start to believe that nothing good could come of it. It's when you hit that low, that frantic part of your life when you feel everything is collapsing, that you have to make a choice. You can either succumb to the visions and descend into darkness, or fight them with every ounce of your soul. I have been to that precipice many times and have chosen to come back from the shadow."

"But how can you come back from that edge? How can you drive away the dreams?" Jared questioned.

"It's hard at first. You have to focus on your life and what is truly important. If you can keep your focus, you can make yourself believe that what you are seeing is just a dream. Not every dream has to mean something. Most of the time, there is no significance to any of your nightmares. You have to focus on who you are and shrug off the nightmares. If you cannot, you will be ruled by the dream world, moving ever closer to madness until you finally snap and succumb to the dark visions."

"It sounds easy when you explain it. I just wish I could have a single night's sleep without such dark visions. I wish I could just sleep without waking up every hour!" Though exasperated, young Jared was beginning to feel better, but worry was still apparent in his features.

"There is nothing easy about it. You have to make a conscious choice and fight with all of your might to keep your fragile sanity intact. Trust in your instincts, remember who you are and cling to those convictions. If you can do that, your nightmares will be bearable. Never pleasant, just bearable."

Her words had struck a chord with the tribal. He considered her wisdom and nodded. Something inside him changed. The perception of sliding downwards was now transformed into an upward spiral driven by firm ideals and his feeling of self. Jared would cling to his convictions and avoid the slide towards madness. A smile graced his lips. "Thanks," he said in appreciation.

She smiled back, blue eyes twinkling in the dark night. "I am glad I could help."

Sighing, Jared looked off into the darkness. It was still many hours before sunrise and he was still responsible for keeping watch over the camp. He felt better for the moment about his dreams, and his mind began to wander. It had been many months since the tribal had had a heart-to-heart talk with Mineera, and it felt refreshing. Already having breached the emotional waters, Jared was no longer hesitant to continue the conversation. Changing the tone of their discourse, the tribal took the opportunity to quiz Mineera about her strange powers.

"What's it like to be like you?"

Not sure what he was after, she responded hesitantly. "What do you mean?"

"You know, your abilities. What is it like?"

Taken aback by his frankness, she formulated her thoughts before replying. "Each psychic has different abilities. No two psychics are alike, but sometimes they share talents. I can feel emotion in others, but it's fragmented. Think of a crowded room where everyone is talking at the same time. You can hear bits and pieces of every conversation, but you can only focus on one. When you focus on one conversation, it's easy to get distracted by the others around you, and you then start to listen to another conversation. Being able to see the things I see is like a crowded room where I cannot *choose* which conversation to listen to. Fragments of sound, images, and emotion flow from a person, and I can sense this. I cannot choose what I see or what will be revealed. I could see a person's childhood or feel their current emotion. I could see how they are going to die or glimpse a long-lost loved one. It's extremely potent but very difficult to control. You never know what you will see or feel."

"How do you keep it all straight? How do you know what you are feeling and how do you keep it separated from what *other* people are feeling? It seems like it would be very easy to get confused." The question was profound. Mineera had struggled with Jared's question her entire life.

"The hardest thing I have ever done was finding myself, my true self. It's so easy to lose yourself when you are intuitive. It's easy to let the full gambit of emotions, other people's emotions, rule your life. If you give in to the voices and emotions that are not your own, you will truly give yourself over to madness. To survive these voices and visions, I make sure that I change my frame of thought away from someone else's existence as quickly as possible. If I don't, I start to lose myself and my ability to isolate my own set of emotions. Getting lost in the world of others is the path to madness. The ability to feel anything through others and see other people's realities has given me more than a dozen lifetimes of memories and experiences. Though I have to battle losing my mind and self, the ability to see anything, feel anything, is a real gift. After knowing what I know and having seen what I have seen, I would never give up my gift of intuition."

Elated by Mineera's revelations about herself, Jared smiled and let his imagination wander. Thinking about her abilities and wild life, he felt lucky to have met her. What an amazing life it must be, feeling and seeing a host of events that comprised the entire human experience.

Nodding in appreciation at her openness, Jared smiled and suddenly felt restless. While they were conversing, no one was watching over the camp. Feeling a little anxious to get back to his guard duties, he stood and prepared to patrol the camp. "I am going to walk the perimeter. You look tired. Why don't you get some sleep?"

Nodding in agreement, Mineera lay back, pulling the blanket over her.

Hesitating, Jared looked at her and spoke earnestly. "Thanks."

Returning his gaze, Mineera smiled. "Thank you. For the first time in a long time, I feel like I belong."

Jared grabbed his submachine gun and moved off to the edge of camp. Sneaking through the rubble, the tribal warrior focused on the nightscape around him. The waves of the ocean crashed against the ruins. White spray rained down as Jared moved nearer to the sea wall.

An ancient concrete structure, once a bank, had collapsed and was now holding back the relentless pounding of the surf. With a feeling of calm, Jared crouched down atop the fallen remains of the shattered bank. Soaking up his surroundings, he let his senses overwhelm his consciousness.

Jared huddled against a fallen wall, watching the violent surge of the black waves. Rubbing his shoulders, the youth let his submachine gun rest on the ground. All was clear around the campsite. No enemies were lying in wait. No violent assassins were plotting their demise. With a smile, Jared rested upon the ruined bank building, watching the ocean waves crash against the ruins until the sun rose over the battered world once more.

Chapter 21
Toxic Evolution

"This is it!" Tani said excitedly. "I can feel it!"

The small boat bounced over the ocean waves, pressing deeper into the ruins of the once-mighty city. Ages ago, a powerful earthquake had reduced the western portion of the city to a fractured mass of jagged earth. The majority of the city was now submerged beneath the violent waves of the sea. The remaining portion was a splintered chain of ruins comprised of hundreds of islands, known as the Shattered Isles.

Nova 7 had spent the last several weeks scouring the islands, stumbling through seemingly endless ruins in search of functional water purifier parts, parts that would hopefully end the war between the rat tribe and the Slumlanders. Using the clues provided by the failed Slumlander expedition into the Shattered Isles, Nova 7 was attempting to find a needle in a haystack.

"Don't get your hopes up, Tani. We have been out here a long time and we have at least a dozen more islands to scour," Banion replied, gripping the railing of the boat. He was standing at the bow of the ship, watching for submerged wreckage. The ship bounced downward and a spray of sea foam and salt water came crashing down on the front of the boat, showering the dour rancher with a blast of cold water.

"Damn it, Tani!" He turned to give the tribal youth a look of alarm. "Watch what you're doing!"

Tani rolled his eyes with a look of mischief behind his wire-rim glasses. The scholar gripped the wheel of the boat tightly with both hands as Jared began to chuckle. The irate gunman soaked on the bow of the ship was too comical to ignore.

The small boat lumbered onward, deeper into the flooded ruins. Mineera was sitting on the back of the ship, soaking up the warm rays of the sun and eyeing the ruins with suspicion. Long tendrils of former glory rested in the crashing waves. An apartment building came into view on the right side of the boat. The rotting corpse of the building, with its curved rebar and waterlogged bricks, was a forlorn sight to the psychic. Most of the building was in ruins and only a small portion of the top floor still remained intact. The majority of the building had been claimed by the powerful tide over the centuries. As they passed, Mineera perceived an eerie sense of forgotten whispers rolling from the building. In ancient times, the apartment building had been a place of life and activity. Now only whispers of sorrow could be heard amongst the moldy carpet and rusted appliances.

Shaking off the haunting voices, Mineera stood up and moved over to the wheelhouse of the boat. Still steering, Tani smiled as she entered. "This is it!" the scholar boasted again. "See all those industrial buildings and factories?" Pointing his stumpy, mangled hand, Tani attempted to motion toward their destination, a large island covered with an abandoned industrial complex. "If we can find water purifier parts, it will be there."

"Hard to port!" Banion yelled as the boat charged through the water. As Tani rolled the wheel, the boat lurched and maneuvered around the submerged girders and debris. Tani skillfully dodged the sunken menace with grace and ease. Over the course of the last several weeks, the young tribal genius had been gradually turning into a wonderful captain.

"Missed another one," Jared said with a mild look of anxiety on his face. Seeing the peril slip away behind the boat, the tribal warrior began to smile. "Hey, Banion!" he shouted. "That was close! Want to borrow Tani's glasses?"

"Real funny, you little bastard!" he roared back with a scowl. "Seeing this junk in the water is harder than you think."

Shaking their heads, Mineera, Tani and Jared eyed Banion with an amused rumble of happiness building in their bellies.

"Slow it down a bit," Mineera urged Tani. "We are getting close."

The tiny boat slowed as Tani cut back on the engines, allowing the vessel to coast closer to the island. The cliffs rising above the waves were monstrous, towering over fifty feet above the ocean. Sewer pipes, as well as subterranean phone and power cables, erupted from the cliff wall. It looked as if a giant knife had cut away the earth around the island, leaving behind a cross section of the island itself. As the boat neared the base of the island and passed behind the rock face, the sun disappeared, bathing the team in a chilling shadow.

The foursome stared up in awe. At their current position, Nova 7 was five stories below the top edge of the island. As the boat bobbed up and down, the team looked at each other with wide-eyed wonder.

"How do we get up there?" Jared asked, trying to imagine a way up the sheer rock face.

"Not sure." Banion shook his head while surveying what lay ahead.

"We have rope, but no way to secure it." Tani shook his head as well, feeling intimidated by the sheer cliff wall.

And so they stewed for a few moments in silence, as each member tried to conjure a solution to the problem at hand. Indecision was rampant until Mineera breeched the silence.

"What are those pipes? You see that? Is that a sewage pipe? It's half submerged in the waves. I bet the pipes could lead back up to the surface." With a methodical gaze, she followed the sewer pipe with her bright blue eyes.

The rest of the team looked at each other, evaluating the psychic's plan.

Finally, Banion nodded in agreement. "At this point, it's our only way to get up there. We can't climb up the cliff. What do you think, you two little gutter rats?" Both tribals nodded in agreement, and the next course of action was set in motion; the team would travel into the partially submerged sewer pipe in hopes of reaching the top of the island.

Gently pushing the throttle, Tani steered the boat towards the base of the sewage pipe. The team grabbed their gear and prepared to explore the ancient waste tunnel. Jared leapt off the front of the boat with a rope in hand. The agile tribal landed on a large boulder and secured the ship, tethering it to the rubble. The rest of the team followed him down, eyeing the entrance to the flooded underworld beyond.

The waves of the sea crashed into the open pipe. Only its upper section remained above the waves. Much of the sewer passage was submerged below the water, bathed in shadow. With a concerned look, Banion turned on the flashlight at the end of his assault rifle and jumped into the water. He shivered as the cold waves bit his skin. Placing the rifle against his shoulder, he directed the beam of his flashlight to penetrate the dark tunnel.

Seaweed and other forms of ocean flora were growing in the murky tunnel. The air was humid and an oppressive smell emanated from the passage. Something which smelled like spoiled fish was rotting within. Standing in waist-deep water, Banion took a step forward. The bottom of the pipe was slippery, and his combat boots crushed the slimy seaweed coating the tunnel's surface. Shaking his head, Banion turned around, eyeing the tribals. "This might be a tale to leave out when you are sitting around a campfire with your grandkids."

The two tribals smiled, balancing their submachine guns against their shoulders. The dull light from their barrel-mounted flashlights barely penetrated the sinister blackness ahead.

Slogging into the ancient sewer, Nova 7 trudged very carefully. The pipe was clogged with rotting ocean plants. Crabs clung to the walls and ceiling, skittering around like a swarm of spiders. In the dim light, the constant movement of the tiny shelled creatures was truly unsettling. The skittering sea creatures were so densely packed together that their legs were intertwined like a matrix of thick spines. Tiny eye stalks moved back and forth, tracking the intruders carefully. The eyes of the tiny creatures shot from the tops of their shells, almost resembling antennae. As the human invaders pressed onward into the flooded tunnel, the small cowardly creatures clambered for cover.

A dense fungal growth hung down from the ceiling, its foul-smelling tendrils brushing across their skin as they passed by. The slick, slimy vines were a cloying nightmare. The densely packed plant life only allowed slow movement and a very minimal line of sight. As they passed through the fungus, dozens of crabs skittered onto the companions. It was truly unnerving feeling the tiny legs of the creatures brush against their flesh as the slimy fungus brushed across their skin.

Banion ripped entire handfuls of the fungus free from the ceiling, throwing it down with an aggressive gesture. The tribals also made their way through by pulling more of it down. Soon the water was a clotted mess, filled with the slimy fungus and seaweed. Mineera passed through the tendrils with her hands gracefully folded, ignoring their creepy surroundings.

After about thirty yards, the tunnel opened onto a room that served as a junction in the submerged sewer. The room was generous in size, over fifteen yards in both length and width. It was devoid of the strange fungus but was populated with enormous, brightly colored stalks, each roughly the size of a tree. The strange, elongated tubes were nearly twenty feet in length, curving down into the water. Most were bright purple and a few had a twinge of yellow, arranged in oblong spots. The stalks were glistening in the dim light, almost like the skin on a wet frog. The room was strangely empty; the swarm of crabs and other skittering creatures from the passage beyond chose to keep their distance.

A ladder extended down from an opening in the ceiling, a manhole through which warm sunlight streamed from the world above, illuminating the chamber in a dull glow. The ladder ended at the center of the room, a tangible escape from the tunnels and a clear route to the top of the island.

Banion raised his fist, alerting the rest of his team to stop moving. Each of them scanned the room warily. Though the strange purple stalks seemed stationary, no one in Nova 7 was too keen on advancing further.

"What the hell are those *things?*" Banion asked in a whisper.

The other team members were silent. Jared pressed the submachine gun against his shoulder, targeting the closest brightly colored stalk. Tani allowed his weapon to drop as he viewed the strange objects. Mineera extended her psychic sensors, but could not feel a thing.

"I have never seen such things in any of my books. It's probably some sort of worm, or maybe even a plant," the young scholar whispered, eyeing the enormous purple tube-like growths with a scowl.

Jared scanned the water in the room, looking for more

information. As his flashlight passed along the water's edge, the white glistening of pearl-white objects grew visible in the water. The tribal took a step forward, disturbing the water. In the dim light, he could barely make out what the submerged white objects were. With an inquisitive look, Jared crouched and reached out to grab one of them. Trembling slightly, his fingers closed around something long and dense. As it emerged from the water, Jared's fist was revealed to be clutching a thigh bone, a human thigh bone, glimmering white as if it had been polished. The rest of the team stared at the human remains with a growing tremor of fear. Moving close to Jared, Tani went bobbing in the water and emerged with a human skull. Gasping, the tribal dropped the skull in the water with a splash, backing away in fear.

"Looks like we found the Slumlander team," Mineera said as her hand disappeared into her blue robes, emerging with a serrated dagger. Her blue eyes darted back and forth, watching the strange, purplish-yellow, slimy stalks. "Let's get up that ladder right now."

A sudden flash of movement caught their eye. The slimy stalks were stirring in the water. With a yelp, Jared stumbled, collapsing into the filthy slime. Something had a hold of the tribal warrior and was pulling him across the room, toward the strange tubes. Coughing, Jared's head emerged, gasping for air. His gun had been lost in the water, forcing him to fumble for his blade. The struggling tribal was defenseless, and was being dragged across the room at a frightening pace.

Like a mass of writhing maggots, the brightly colored stalks began to flail about. Movement was everywhere in the room. Other stalks, hidden in the mire, burst forth, seeking victims.

"Hit 'em!" Banion yelled, leveling his weapon. Pulling the trigger, he opened fire, initiating a frightening assault. The bullets tore forward, slamming into the nearest purple stalk. The volley of bullets was precise, tearing massive holes in the unusual creature. A splash of thick, black ooze erupted as the beast thrashed about. With a flurry of movement, the mighty creature emerged from the water with frightening speed. It was much larger than its brightly colored stalks had indicated, and Nova 7 was immediately intimidated by this aberration of nature.

The head of the creature was a massive opening, surrounded by hundreds of pale white tentacles. Standing upright, the mutant sea anemone hurled its tentacles at the fallen tribal warrior who was huddled only a few feet away. The sticky tendrils clutched the tribal, lifting him into the air. Held in the pale white strands, Jared was helpless, his limbs entrapped and entangled. The mighty creature opened its mouth, exposing the darkness of its digestive tract.

"Aim at the base, Tani!" Banion yelled, still firing at the creature, directing his gunfire into the knotted mass of fibrous flesh at the bottom of the creature.

The tribal scholar panicked and pulled the trigger on his submachine gun, causing the weapon to erupt in a hail of lead. The terrified Tani was unable to control the weapon. The barrel of the gun flew upwards from the recoil and he quickly drained the clip, ramming most of the bullets into the ceiling. Not a single bullet hit its mark.

Mineera, heeding Banion's instructions, was focusing on the base of the monster. As she extended her hand in the air, a spark of fire rolled from her elbow, coalescing into a potent, fiery blast. The white searing beam of psychic energy lit the tunnel with a blast as powerful as the sun. The flash was so intense that the purple stalk of the monster crackled and popped as the heat incinerated its flesh. As the purple color turned to a charred mass of blackened flesh, the creature writhed and its tentacles loosened. Jared fell from the creature's grasp and tumbled head-first into the murky water.

Trying to stand, the tribal was beset by yet another monster. Balancing on one knee, Jared gasped for breath as he tried to draw his sword. As his fingers closed around the mighty Scar Blade, another set of tentacles grappled with the youth. The mighty blue sword was but a few inches from the scabbard when Jared was dragged underwater once more. The Scar Blade slipped away from his grasp, filling Jared with panic as he was dragged deeper.

Banion moved forward with weapon in hand, firing short bursts at the other creature now pulling Jared toward its open mouth. Ooze erupted as the bullets mangled its stalk. As the monster died in Banion's crosshairs, two more mutants emerged from the water, each throwing their tentacles around Jared. The two enormous sea anemones tugged and pulled at him as if he were a wishbone, playing a hungry tug of war.

"Damn it, do something!" Jared screamed.

Tani was shaking, frantically trying to reload his weapon. A fresh magazine of ammunition crashed into the water and the scholar hurriedly searched for it on his hands and knees.

Mineera was calm, trying to focus on the fight. Another creature erupted from the water, springing its tentacles in her direction. Now in a battle of her own, Mineera blasted away at the beast confronting her. Another flurry of psychic, holy flame roared forth, striking the monster and turning it into a singed wreck of burning flesh.

Banion's weapon jammed. Cursing, the mercenary fumbled with the firing chamber. A smoking shell casing was jammed in the chamber, and the legendary hero knew it would take a considerable amount of time to fix the weapon. Abandoning the assault rifle, Banion caught sight of the mighty Scar Blade and grasped it with both hands. Bellowing in rage, the rancher charged and started hacking away at one of the beasts. His clumsy attacks were rapid, but inflicted little damage due to his lack of skill with a blade.

With an echoing splash, one of the creatures collapsed into the water, its smoking body crashing down with a plume of rising steam. Mineera jumped over the body, charging forward into the fray. Seeing Tani still gripped with panic, Mineera rushed towards him and yelled at him. "You're a demolitions expert! Drop the gun and blow the damn thing up!"

A flash of recognition filled the tribal's face. He acknowledged her command and grabbed a hand grenade from his belt. Mineera had drawn dangerously close to the confrontation, trying to focus on slaying another beast. Tani followed and pulled the pin from the grenade. Carefully watching the open maw of the beast trying to devour his friend, Tani held his breath and aimed, intent on making a bull's-eye shot. With a well-aimed toss, Tani launched the explosive into the open mouth of the mutant anemone. A split second passed and the weapon detonated. Fragments of flesh and metal exploded from the head of the creature. A spray of black gore rained down as the mangled monster died. With a loud splash, the dead monstrosity crashed into the pool, its blood further staining the murky water.

With only one beast remaining, the fight was on. The final monster was very close to ingesting Jared by passing him into its open maw. A blast of white light struck the creature. Banion clumsily hacked away with the Scar Blade. Tani was grabbing his friend's legs, battling to keep the monster from eating him alive.

"Jam the blade in and then rotate it around in the wound!" the tribal warrior yelled, covering his face in a protective gesture as the stench of the creature's gut filled the air.

Banion listened to Jared's advice, jamming the blue steel blade into the body of the monster. With a yell of determination, the frantic gunfighter spun the sword around like a drill inside the monster's body. The ensuing wound was truly gory. Black blood erupted, spewing forth as the blue blade ground deeper. Swaying back and forth, the mutant sea monster was in mortal trouble. Another blast of holy fire struck the monster, singeing its body in a frightening display of unleashed psychic energy. The damage wrought by Nova 7 was terrible. Unable to survive the assault, the final mutant beast died.

The calm after the battle was eerie. Where conflict had raged, now there was only silence.

Jared staggered to his feet with wide eyed fear still lingering on his face. Pulling the Scar Blade from Banion's hands, the warrior spun around in a circle, looking for more creatures. None could be seen, but the sound of something splashing through the stagnant water down the southern tunnel made the urgency of escape all the more real. Recovering their gear, the rattled team of mercenaries rushed toward the ladder in the middle of the room. With brilliant sunlight pouring from the opening in the ceiling, the exit was an appealing alternative to the sinister dark network of sewer tunnels. Climbing upward, Nova 7 quickly escaped the abandoned sewer tunnels and its sinister inhabitants.

Chapter 22
Ruins of Industry

The sickly sweet smell washed over the team.   After climbing a ladder to escape the treacherous underground tunnels, the members of Nova 7 now found themselves standing atop a giant mound of garbage, gazing out in awe at the ruins surrounding them. All around them, like dunes of sand, were enormous piles of refuse. Ancient mounds of trash pressed on in every direction, and the stench was overpowering.

"This is wonderful," Tani mouthed with mounting sarcasm. "With all the wonders of the ancients, the last thing I ever wanted to see is a landfill."

Jared chuckled as he eyed the enormous piles of garbage.

"It has been ages since the fall of man. I cannot believe all of this refuse is *still* here." Mineera shook her head in dismay.

"Let's just get through this mess," Banion said, taking a step forward into the shifting mass of garbage. The others reluctantly followed, pressing deeper into the ancient landfill.

At the bottom of one of the mighty mounds of trash was a valley of sorts. Over the years, residue from chemical containers in the landfill had mixed with rainwater, creating a shimmering pool of rainbow colored liquid. A sickly sweet smell was emanating from the toxic pool of vile liquid. The middle of the pool was turbulent. Gasses from the bowels of the landfill were pushing upwards to the surface, bubbling through the sludge, infusing the toxic morass with additional chemical compounds. All around the perimeter of the pond stood strange plants, their gnarled roots disappearing into the vat of noxious chemicals. Standing tall, the spiky plants rose several feet in the air. At the end of each stalk was a pod. A clear, mucous-like substance dripped slowly from each pod back into the rainbow colored pool of toxic chemicals. Dead birds were suspended in the viscous, rainbow colored pit of chemicals. The acidic solution was slowly dissolving their carcasses.

"It's an enormous bio-reactor," Tani mouthed, observing the strange plants growing in the sickening pit. "Over the centuries, these plants have developed an ability to harvest nutrients from the chemical sludge. It must have taken hundreds of years for such mutations to take hold. This is a true wonder of evolution. Without a stable supply of normal nutrients from soil, these plants have developed metabolisms to handle the substances in this chemical vat."

"Yeah, real neat, bookworm. Just don't get too close. I'm not pulling you out of there if you fall in that stuff," Banion snorted, taunting the tribal with a crazed look in his eye.

Tani nodded in agreement, taking a step backward up the pile of trash, putting some distance between him and the noxious pit of chemicals.

"The ancients were truly masters of evolution," he said, shaking his head as a strange theory filled his mind. The other members of Nova 7 had no idea what the tribal youth was talking about. Reserving judgment, they viewed the young genius with silence.

"No, really, think about it. In ancient times, garbage was deposited in this landfill. Nutrients from dead plant matter and remnants of vegetables and fruit rotted here, and the chemicals filtered downward through the garbage, carried by rain water. Other substances, such as battery acid, cleaning supplies, and other toxic chemicals, mixed with the nutrients of decaying food. The result is a concentrated supply of chemical energy, the likes of which the world has never seen. Raw materials capable of creating life itself are housed in this landfill."

Pointing at the bubbles percolating up through the rainbow colored sludge, the tribal continued his strange theory.

"The bottom of the landfill must be filled with thousands of micro-organisms consuming the garbage. The result of such a metabolic feast is a torrent of noxious gases, by-products of the bacteria consuming the trash. This leads to a constant flow of gas that rises through the garbage, further infusing the layers with additional chemical compounds. What we are looking at is a huge life generator, a place where any chemical imaginable can serve as fuel for mutated life forms."

The rest of the team simply looked at Tani with raised eyebrows. Undaunted by their silence, the tribal scholar continued.

"The creatures in the sewer passage below are beginning to make more sense. As rainwater pushes downward into the landfill, it must pick up chemical residue. Some of this chemical stew must penetrate deep into the earth, where it's deposited into the tunnels below. The chemicals must have been strong enough to induce mutation. We best be on our guard. With so much mutagenic material on this island, and an immense supply of chemical materials, there is no telling what other forms of life we will encounter," Tani concluded earnestly, pushing the glasses up his nose.

"I don't sense any intelligent life out here. It will be hard to sense an ambush by mutants," Mineera said softly, looking around at the mounds of plastic and hi-tech polymer that had resisted the passage of time and were still littering the landfill.

"Yeah, I have had my fill of mutant attacks for one day," Jared replied, remembering the last attack in the sewer pipes below.

Banion moved forward, sliding down the piles of trash. He lost his footing on several occasions and had to steady himself. Working their way through the mounds of garbage, Nova 7 followed the gunfighter, skirting other toxic pits of sludge. Within minutes, the mercenary team had liberated themselves from the ancient landfill.

A collection of steel piping and enormous holding tanks came into view. Only mildly corroded even after all the centuries since the fall of man, the complex was mostly intact. In ancient times, crude oil was processed in the site, using enormous distillation processors to yield precious fuels such as gasoline. Now the abandoned oil refinery was a lonely testament to a world gone mad. Their elation at locating such technology still intact made each member of the team feel hopeful regarding their goal of locating water purifier parts.

Pressing onward into the lonely refinery, the team members

heard the sound of birds crying. Turning toward the source of sound, they caught sight of an enormous series of steel buildings rising high into the air. Upon the monstrous structure were a host of nests, with thousands upon thousands of ocean birds dotting the top of the refinery towers. The nests were made of garbage from the landfill and clung to precarious perches. Grotesquely painted in animal feces, the top of each tower was stained brilliant white. The bird excrement had permanently covered the steel with a thick layer of filth. As the mercenary team moved closer, cries of alarm rose from the seagulls. Flapping their wings intently, the hundreds of white gulls screeched repeatedly at the group's approach.

"The purifier parts are here. I can feel it," Tani said triumphantly, excited to have finally reached the oil refinery that the team had seen from the ocean below. "With so much industry on this island, I *know* we can find what we are looking for."

"What is our best course of action, Tani? Where do you think purifier parts would be located?" Banion eyed the tribal scholar expectantly, prepared to trust his opinion on the matter.

"Over there, past the refinery. It looks like some sort of fabrication plant." Tani gestured with his stumpy hand, pointing out the general direction. Beyond the ruined oil refinery was another industrial complex.

The mercenary team pressed deeper into the abandoned oil refinery. As they moved on, the gulls on the towers continued to cry in alarm. Many took flight and began to circle. Undaunted, Banion led Nova 7 into the oil refinery complex. With mounting agitation, the gulls began to descend lower and lower, crying ever more desperately, trying to drive back the human invaders. Several came frighteningly close, making each of them jump or duck to escape the dive-bombing birds. The angry gulls tried relentlessly to drive Nova 7 back. It was becoming comical, ducking to evade the menacing birds. Banion picked up the pace and started to walk quickly, wanting to get away from the obnoxious avian attackers. As he pressed into the center of the refinery, the bird attacks stopped. With squawks of panic, the gulls retreated, flying away from the middle of the refinery. They perched atop the stacks, eyeing Nova 7 with anxious gazes. All the birds in the vicinity suddenly became deathly quiet.

Thousands of tiny eyes stared at them. Exposed and in the

open, standing in the center of the refinery with confused looks, Nova 7 eyed each other with rising fear. Spinning around, Banion watched the birds with suspicion. Jared, too, looked up at the gulls. The birds seemed haunted, as if anticipating some horrid event.

"Let's get the hell out of here," Banion said in panic.

Slurp! The sound of something splattering against the concrete filled their ears. Spinning around, the team looked down at an object which had landed near them. It was small and round, about five inches in diameter. The blob of flesh was clear and transparent, with fibrous tendrils implanted within the gelatinous matrix. It quivered upon the ground like a trembling mound of jelly. A slurping sound erupted from the sinister blob as it quivered upon the ground. It seemed to be confused, moving back and forth, side to side, as if trying to determine its next course of action. Each member of Nova 7 held their breath and eyed the blob with anxiety.

As a moment of silence gripped the team, Jared looked up. Above the mercenary team were a series of gantries, piping, and catwalks. Clinging to the bottom of the piping and walkways were thousands more of the strange blobs of jelly, hanging upside down. Slowly, the blobs were moving to perfectly position themselves over Nova 7.

"Run!" Jared yelled, bolting past Banion.

The gunfighter stared at the tribal youth as if he were a madman. Mineera looked around frantically, her senses trying to probe the source of danger. The tribal scholar, not accustomed to Jared reacting in such a manner, ran after him, knowing full well his friend from Scarskin had a good reason to escape the oil refinery.

Banion and Mineera stood dumfounded; each had drawn their weapons and now spun around in circles looking for the threat. The gulls eyed the team and cried in panic. With heart pounding, not knowing what to think, Banion started to run after the tribals. Mineera gave chase as well, and none too soon.

Like the rain they came. Leaping off the piping and gantries, thousands of the gelatinous blobs launched downward, trying to land upon the members of Nova 7. Banion and Mineera were in the middle of a storm of Jellyslugs. All around them, the blobs hit the concrete with a splatter and a slurp. Jumping from left to right, Banion leapt over the blobs.

Mineera tried desperately to shield herself. Using her

psychic powers on the fly was tough. Barely able to concentrate while moving, the powerful psychic managed to erect a barrier of spiritual energy. The white psychic barrier surrounded her as she stumbled through the middle of the refinery. The Jellyslugs slammed into the shield, sliding down with a slurp. Several dozen struck Mineera's potent psychic barrier, allowing her a brief reprieve from the attack. With dizziness filling her as a result of using so much spiritual energy, the psychic barrier began to fail. Running full speed, fighting the fainting spell tugging at her consciousness, she managed to escape the hungry sting of the Jellyslugs. Banion was not so lucky.

With a loud splatter, a Jellyslug landed on Banion's shoulder. A burst of liquid erupted from the gelatinous creature. Soaking immediately through his long coat, the acidic liquid began to dissolve his flesh. A searing blast of pain erupted from his shoulder. The blast of acid was so strong that the skin on his shoulder began to dissolve within a matter of seconds. Gritting his teeth from the pain, Banion drew his knife and stabbed the creature. With a flip of his wrist, he flung the slimy blob off his shoulder. It hit the ground and formed a puddle, its body punctured, spilling its contents. Wincing, Banion charged onward, toward the tribals.

Jared and Tani had stopped, now clear of the overhanging pipes and gantries, watching helplessly as Mineera and Banion tried to escape the thousands of Jellyslugs raining upon them. Dodging and leaping, Banion and Mineera managed to reach the outskirts of the refinery.

Gasping in pain, Banion clutched his shoulder. His long coat had an enormous hole in it. The acid had eaten away the fabric of his clothing. Blood was streaming from the wound where his flesh had dissolved. A sickly sweet smell filled the air as the acid continued to burn his shoulder.

"Wash it off," Tani said, grabbing his canteen and pouring water into the wound, cleansing the acid from it.

Gritting his teeth, Banion breathed heavily, closing his eyes and hoping the pain would stop. "Thanks, Tani."

"Good thing they can't move fast," Jared said, still watching the strange creatures. The tribal's words were short lived.

Like a mass of rolling jelly, the blobs began to lurch forward with frightening speed. As they moved, two became one, and the

Jellyslugs merged together. Within a few seconds, hundreds of the blobs had joined together into an enormous mass of jelly. Gasping in terror, Nova 7 took a step backwards as the enormous blob launched itself into the air. The mighty blob rose several stories high in the air. As the creature composed of many reached the apex of its jump, the ball of jelly exploded. The gigantic mass of Jellyslugs burst apart and each individual component was launched more than thirty yards, spraying the surrounding area with hundreds of individual slugs. The attack was so sudden and unexpected that Nova 7 was left unprepared.

Dozens of the smaller slugs landed between the team members, startling them. With a slurp, the tiny Jellyslugs launched at each member of Nova 7. Jared was the best prepared, his reflexes allowing him just enough time to draw the mighty Scar Blade. With a heavy swing, Jared managed to strike a slug leaping at his face. The blade hit the creature and cleaved it in two, scattering the acidic, gelatinous contents around the battleground.

Banion had had enough. He drew his powerful silver revolver and began to take shots at the creatures. Being a master marksman, the gunfighter gained the upper hand by killing several of the creatures with expertly placed shots. With a splatter, they broke apart as the bullets struck them.

A white ray of burning light erupted from Mineera's outstretched fingers. The burning stream of energy leapt from her hand, searing and burning its way through the host of slugs.

Tani, who had been hit by two slugs, was emitting a primitive, terrified wail. Using his knife, the wily scholar cleaned the slugs off before they unloaded their acid cargo upon his body.

As Nova 7 fought against the host of Jellyslugs, thousands more caught the scent of blood and warm flesh. Drawn to the conflict, they streamed in, driven only by instinct and hunger. Fighting a losing battle, Nova 7 knew escape would be impossible. They were already being surrounded and blocked by a wall at their back, and hope was failing. The team had been cornered, and thousands of Jellyslugs were rushing to the scene, slurping and leaping, many forming into enormous collective blobs of slime.

"Someone think of something fast. We are being overrun!" Banion screamed, now holding his assault rifle, gunning down more Jellyslugs with staggeringly well placed shots.

"No ideas here," Jared yelled back, swinging his sword like a baseball bat and splattering the blobs.

It took a brief moment for Tani to formulate a plan. Looking around the scene, he noticed dozens of holding tanks, filled with highly explosive fluids produced in the refineries long ago. "I hope this works..." Tani whispered, the cogs in his brain feverishly grinding away.

Pulling his submachine gun, the wily scholar took aim at the holding tanks. With a shaky motion, he opened fire, sending several bullets into the ancient petroleum equipment. The bullets barely inflicted any damage and only a few small holes opened. Clear flammable liquids began to trickle from the small holes, but he needed more damage to fully execute his explosive vision.

Drawing dangerously near, several hundred more Jellyslugs were only a few moments from overwhelming the mercenary team.

"Hit that tank with everything you've got!" Tani yelled, firing more bullets into the tank. Jared drew his own submachine gun and opened fire. Banion blasted away with his assault rifle.

Seeing the potential for an enormous explosion, Mineera was concocting a plan of her own. Knowing full well that the difference between life and death would be a matter of seconds, she dropped into a trance with the intent of shielding Nova 7 from a potential explosion.

The gunfire from all of the automatic weapons tore at the tank. Focused on roughly the same area, the bullets managed to tear a gaping, jagged wound in the side of the tank. Hundreds of gallons of volatile petroleum distillates poured through. The advancing Jellyslugs were showered and submerged in the explosive liquid.

Yelling in pain, a Jellyslug released its acidic cargo on Jared's leg. The tribal dropped his gun, prodding the blob off his leg with his sword. "Do something, fast! We are being overrun!"

With a resolute look, Tani grabbed a grenade. Pulling the pin, he ignored Jared's screams, the slurping of the Jellyslugs, and the strong smell of gasoline wafting through the air. Focusing on survival, the wily scholar lobbed the grenade into the pool of explosive gasoline pouring out of the holding tank. "Now, Mineera!" Tani yelled as the grenade landed in the enormous pool of explosive liquid.

As she threw her hands forward, the white barrier of psychic energy erupted from her fingertips, and just in time. The white glow surrounded Nova 7 as the grenade detonated, igniting the volatile liquids. With a bright flash, the refinery was lit by an explosion. Orange flames leapt high into the air and a shockwave of energy rolled forward. Bracing for the impact, the members of Nova 7 ducked down, not sure if Mineera was strong enough to hold back such an explosion. As the wave of flame rushed forward to engulf the team, Mineera held her concentration. As the flames met the holy glow of her psychic shield, it held strong, protecting them from the blast.

The horde of Jellyslugs were incinerated, utterly burned and scoured by the explosion. Where only a few minutes ago thousands of such creatures were on the attack, the surviving Jellyslugs were quickly driven back by the heat and flame. Motivated by instinct, the primitive blobs of quivering jelly retreated. With most having either retreated or been destroyed by the blast, the remaining dozen or so slugs were easily dispatched by Nova 7.

As the damage from the initial blast subsided, more flames rose throughout the refinery. The explosion had sparked other fires, fires that were now lighting other holding tanks, causing a massive chain reaction of explosions. Watching the rising plume of black smoke erupt from the enormous flames, the team was relieved that they had survived the bizarre ambush.

"Damn good idea, Tripwire!" Banion yelled in glee, smiling at Tani, watching the mass of Jellyslugs wriggle in the burning pool of gasoline. Wild with excitement, Banion grasped Mineera's arm with both hands, smiling broadly. "Great job, Mineera. Amazing, the things we can do when we stop acting like selfish bastards and act as a team."

Nodding in agreement, with the rush of adrenaline fading, the team watched in fascination as the gigantic flames consumed the refinery. As the flames ignited the fuel, distant explosions reached their ears. More fuel tanks were exploding under the high temperature and heat from Tani's little fire.

No longer pinned, Nova 7 escaped the refinery, while black smoke rose and curdled into the air and orange flames engulfed the ruins of the ancients. The quest to locate the water purifier parts needed to end the war between the Slumlanders and the rat tribe would soon be fulfilled.

Chapter 23
A Small Victory

Rising clouds of smoke and fire created an enormous thermal updraft of rising destruction. The bright orange flames lit the night as the age-old petroleum reserve continued to burn, a fire that had been raging ever since the mercenary team punctured one of the large holding tanks in a heated battle for survival, earlier in the day. Enormous clouds of black smoke curled into the night sky, blocking the very stars. The eruption of flame and chaos was so grand in scale that it could be seen for miles and miles in every direction. The locals of the region would always remember that night, for when the sun rose the next day, the fate of the Concrete Barrens would be forever changed.

"Don't press so hard!" Jared said shakily, nearly whimpering in pain. Gritting his teeth in anxiety, the tribal warrior peered at Banion as he scraped the wound on his leg with an alcohol soaked rag. Earlier in the day, the tribal had sustained a burn when one of the mutant Jellyslugs unleashed its vile, acidic cargo upon his leg. Since then, the youth had been in pain, his wound already festering. As Banion rubbed back and forth vigorously, it felt as if he were cleaning the wound with sand paper. On the verge of pulling his leg away, the tribal warrior was elated to see him stop. The relief was short lived, and Jared's eyes opened wide as Banion brought out the bottle of whiskey once more. With a sinister grin, he began to dump a copious amount of whiskey onto the open wound.

"Damn it!" Jared yelped in pain, feeling a searing flash of fire erupt on his leg. Slapping him on the shoulder, Banion smiled. He took a couple of swigs from the bottle, then handed it to the tribal warrior. Jared tipped it back into his mouth, glowering at Banion. "You should have let me have some of that whiskey before you cleaned the wound!"

"Nah, it was way too much fun to watch the look on your face." He snorted and threw his cowboy hat down on his sleeping bag. Exhausted from the events of the day, the seasoned mercenary sat down and gave a loud sigh. Watching the orange flames still burning to the northeast made him smile. The smile grew to a chuckle. Turning around, Banion scanned the campsite for Tani.

"Nice work, Tani. That is the biggest damn fire I have ever seen." The whiskey was flowing into his veins and it felt wonderful. The scar on the right side of his face constricted as the smile grew even bigger. The day had been tough, and Banion wanted nothing more than to wash away the events with strong liquor.

Moving toward his two companions near the campsite, Tani felt a little embarrassed with all of the destruction. "Yeah, well... I had to do something."

"*Something?*" Banion snorted again. "You burned an entire island."

"I always knew you liked fire, but come on," Jared added with a smile on his face.

"Yeah... yeah..." The scholar looked around for the bottle of liquor. "Let me have some of that whiskey before you two greedy bastards drink it all." Jared handed the bottle to Tani. Scrunching up his nose, he took several swigs of the strong liquor, throat burning and eyes watering, then tilted the bottle away from his lips with a cough. The liquor was so strong, the scholar almost choked.

"Real smooth, bookworm," Banion said, rubbing his eyes. His vision was growing blurry as the whiskey coursed through his veins. Turning away, he propped his back against a concrete support column and watched the fire blazing at the distant refinery. It was an eerie sight, a never-ending wash of red flames surging and changing form, writhing in the darkness. The smoke would belch from the flames, obscuring the core of the inferno. On several occasions, due to the shifting clouds of smoke, it appeared as if the fire was burning out. But then the wind would suddenly shift, pushing the black smoke away from the flames and almost giving the impression that the fire had exploded once more for a brief second.

Jared and Tani moved over and sat on either side of their leader. The liquor was also taking its toll on the tribals, and they wanted nothing more than to kick back and relax. The stress of the day, which had consisted mostly of a long battle to survive the mutant monstrosities of the island, had been formidable. Tani drank another swig and almost choked on it again. Though getting better, he still had little fortitude when it came to such strong liquor. Rolling his eyes, Jared quickly grabbed the bottle away from his friend and took another swig himself.

The shadows moved. A form emerged from the edge of camp. Wreathed in darkness, Mineera's blue robes ruffled as she moved toward the trio. Each of them squinted, their vision blurry, as she approached.

"All clear," she said softly, sitting down beside them. "I scouted the entire edge of the island. I think we can rest easy tonight."

Banion grabbed the whiskey bottle and promptly handed it to Mineera. She smiled and shook her hand, gesturing that she would pass on the liquor. "Suit yourself, more for me." Banion took another swig and sighed.

The companions sat quietly and watched Tani's 'little fire' rage in the distance. The team had once again escaped death by the skin of their teeth. If not for the tribal scholar's quick thinking, the team would have met a disturbing fate, being dissolved by the nauseating gelatinous creatures.

Drunker by the second, Banion stood up suddenly and held the bottle in his shaky hand. "Let me see *it* one more time, bookworm. Come on, show me those damn water purifier parts."

Tani recovered his backpack. His stumpy fingers probed the interior and returned with a grungy adapter with a host of dials. With a triumphant move, he held the prized water purifier parts high in the air. All of the companions eyed the parts with a sense of accomplishment. After incinerating half the island and escaping the horde of Jellyslugs, the mercenary team had quickly tracked down a water purification processor in a nearby industrial water plant. It took the scholar only a few minutes to liberate the required parts, the very ones which Nova 7 had been seeking over the last few weeks.

The mechanical components were a symbol of hope, a way to stop a war between the Slumlanders and the rat tribe living in the Concrete Barrens. And with an end to the hostilities between the two tribes, the mercenary team could haggle with the Runner for the location of the nuclear warhead. Their goal now one step closer and the team was elated by their small success.

All of the companions were still staring at the water purifier parts when Banion, slurring his words, spoke once more. "I propose a toast. To Tani, the brilliant pyromaniac, may he always light such wonderful fires..." Taking another swig of whiskey, Banion was rapidly beginning to lose his grip. His fortitude with liquor was substantial, considering the fact that he had downed almost a quarter of the bottle within only a few minutes. Lesser men would have been seeing double.

Jared grabbed the bottle and shouted. "To Tani!" Taking a mighty gulp of liquor made his throat burn. He wiped away several drops running down his chin.

The tribal warrior handed the bottle to Mineera. She hesitated a moment, then shook her head, smiling. "To Tani," she said, tipping back the bottle into her open mouth.

Finally, the spiky haired scholar took the bottle. Drinking quickly, he was nearly overpowered by the liquor. Gasping, he coughed yet again, blinking several times as tears rolled down his face. Despite his efforts, he felt the hot, volatile alcohol exit his mouth with a whoosh, dribbling down his chin.

"I never thought we would find them..." Tani coughed again, looking triumphantly at the purifier parts. The reality of their accomplishment was beginning to sink in, even while being utterly intoxicated and deep in a drunken stupor. It was no small feat; the team had risked life and limb recovering the prized mechanical components.

"Me neither," Jared concurred, blinking several times as he watched the orange flames snake into the sky.

"It sure beats the alternative." Mineera spoke in a quiet tone. She was referring to the original deal with King Lavosi, a deal in which Nova 7 was supposed to kill the Slumlander tribe by poisoning them to death with deadly toxins.

"I hope Lavosi deals with us," Tani responded earnestly. "We didn't exactly fulfill our end of the bargain. Hopefully, he'll see the wisdom in our plan."

"There is no reason not to," Jared said. "Once we fix the Slumlanders' water purifier, there is no need for the war to continue. Hopefully, even that twisted mutant rat can see the logic."

"If he doesn't, at least *we* did the right thing." Banion shook his head in disgust as he spoke. "It might be the first time in my life, but I am glad I did the right thing. Killing the innocent to achieve our goal is something I would have embraced without a second thought before meeting all of you. Thanks for bringing me back."

For Banion, it was a profound admission. He had lived his life in constant motion from one broken home to the next, with anger guiding most of his actions. Now, the reckless mercenary was beginning to come to terms with his violent past. It wasn't an event that would forever change his life, but it was a step in the right direction, a first step on a long road to redemption. Where once he would have sold his own soul to see his enemies destroyed, Banion had changed, taking a stand against sliding further toward evil.

The rest of the team looked at each other with a jolt of surprise on their faces. The reckless gun fighter did have a heart, hidden beneath all the scars. It was a great moment for them all. For that brief moment, they all felt empowered.

The moment of calm was shattered by the same man who had brought it about.

"Hand it over!" Banion yelled with a yip. His finger was pointing at the whiskey bottle. The change in emotion was startling but somewhat comical. Complying with his order, Tani handed him the bottle. Taking another swig, Banion began to chuckle. His laughter rolled into the night. The rest of the companions, also feeling the power of the whiskey, stared on in silent amusement. Watching him laugh made their own hearts leap. None of them made a sound; they didn't need to. Banion was lost in his own world, chortling uncontrollably, watching the flames rise into the darkness of the night.

As the sound of their laughing leader rolled through the

night, each member of Nova 7 felt a sense of closure. Having spent over a month scouring the islands for the purifier parts, the mercenary team experienced a feeling of accomplishment. Though the road to the nuclear warhead was still unclear, Nova 7 was one step closer to their insane goal.

As the alcohol filled their minds, a natural wonder began to unfold in front of their eyes, pushing into the night sky with a surprising debut. The night was no ordinary night. Set amidst the rising plume of fire, a white spike of light was revealed.

Seeing the spike of white light, Tani rose to his feet, almost falling over, immersed in the powerful whiskey. Gesturing at the light with his stumpy hand, he exclaimed, "I had completely forgotten!"

Rising amongst the flames, the pure white glow pushed to breach the seclusion of the rising smoke. Just as drunk as his friend, Jared became excited and stood up to view the natural wonder, feeling his heart flutter in anticipation.

A tense moment filled the tribals as they stood watching the lunar event unfold. Breaching the dense tendrils of smoke, the edge of the moon revealed itself over the flaming ruin. Glowing like a smile, the moon pushed upwards into the dark sky. With each passing moment, more of it was revealed. Finally, a perfectly round, brilliant holy glow illuminated the sky. A full moon, pristine and inspiring, hovered in the distance.

For the tribals, the rising of the moon was more than just a natural event; it was a milestone. Having been exiled from their village for twenty full moons was a daunting road, filled with peril. As the full moon shone brightly in the sky, Jared and Tani felt relieved; they were both one step closer to home. Shaking, with a feverish frenzy filling their hearts, the tribals smiled, watching the full moon with glassy, drunken eyes.

"Eleven moons..." Jared spoke in a whisper. The statement had profound meaning for the two. Having survived for eleven full moons outside the safety of the village of Scarskin, the tribals were on the downhill portion of their trek, with only nine full moons keeping them from their loved ones back home.

Seeing the tribals' reaction, Banion stood with shaky legs. Grasping the whiskey bottle, he made a toast to the tribals. "To Jared and Tani – you are both one step closer to home." With a smile, he took a swig of whiskey, then passed the bottle to his tribal companions. As Jared and Tani drank to the milestone of the eleventh moon, the intoxicated gunfighter placed a hand of each of their shoulders, standing between them. Mineera also stood and smiled at the tribals. Stepping before them, she grabbed the whiskey bottle and echoed the toast. "To Jared and Tani." She smiled gracefully as she spoke. Tipping back the whiskey bottle, she felt the volatile liquor roll down her tongue.

The night would always be remembered. The team had done the impossible, finding the purifier parts and surviving mutant attack once more. Standing in the white glow of the fully risen moon, the team watched as orange flames roared into the dark night. The full moon was a powerful omen. It lifted their spirits, gave them hope, and filled them with wonder. With a long road still ahead of them, Nova 7 felt refreshed, and the fire in their souls was rekindled. They had won a small victory that day, bringing each of them one step closer to home.

Chapter 24
Daring Diplomacy

As the mercenary team, with the coveted water purifier parts in tow, entered the perimeter of the Slumlander village, they found the area strangely quiet – a little too quiet. The village, tucked among the ruined buildings of the Concrete Barrens, was enveloped by mist which had rolled in from the nearby sea, choking the ruins with a dense fog which seemed to smother every sound. It was eerie, almost as if the members of Nova 7 were the only beings left alive in the entire settlement.

"I don't like this at all," Banion said, his tone aggressive.

The ruined skyscrapers, which had been filled to the brim with shambling, sickly humans on their first trip to the Slumlander village, now appeared utterly abandoned.

"We are too late." Tani's voice expressed his frustration. From the look of things, it appeared that the war between the Slumlanders and the rat tribe had finally extinguished one of the factions.

"This can't be true," Jared argued, shaking his head in denial. "I can't believe, after everything we've been through, that we are too late."

Though Banion, Jared, and Tani were all disturbed by the turn of events, their reaction paled in comparison to what Mineera was feeling. She was frantic. A scream at the edge of her soul pushed toward the surface. Shaking with despair, the prophet felt betrayed. Fighting for the just and right cause only to have everything end in ruin was inconceivable. With disbelief filling her, she let her senses come to rest. A serene peace washed over her as she stood in the ruins. A tingle greeted her soul. There was something that the team was missing. Hope had not been extinguished. There was still a chance, but that hope was an elusive dream to the sensitive psychic. Closing her eyes, the full scope of the mysterious disappearance of the Slumlander tribe filled her. There was still time. As she opened her eyes, something ahead in the fog caught her attention. Bolting forward, she disappeared into the white darkness.

"Mineera!" the leader of Nova 7 shouted. "Get back here."

She did not respond to the command. Instead, letting her heart lead the way, Mineera charged onward to quell the rising fear clawing at the edge of her soul. There was still hope. There was still a chance to avert massive bloodshed. There was still a chance to stop the war between the Slumlanders and the rat tribe. The answer to the riddle of the quiet in the Concrete Barrens was out there, somewhere in the fog.

The rest of the team had no choice; they picked up the pace and charged after her. Stumbling in her tracks was difficult. The fog had engulfed Mineera. Squinting in the white darkness, each of Mineera's companions strained desperately to catch sight of her. Chasing after her fading footsteps, they halted only a few yards into the cloying white haze, confronted with a curious sight. A child dressed in rags was standing alone in the fog. Her expression was forlorn, and she held a battered doll in her hand, sucking her thumb in an anxious, repetitive gesture.

Not wanting to startle her, Mineera approached cautiously, bending down to look less imposing to the frightened child. She crouched gracefully and looked into the girl's eyes with a gentle gaze. The small girl looked back, blinking several times.

"Are you lost?" Mineera spoke in a soothing tone, placing her hand on the girl's shoulder.

Blinking several more times, the child finally replied. "I am lost." The tears were welling up in her eyes as she spoke.

"Where were you going? Where are your parents?"

"My mommy is in the shelter."

"The shelter?" Banion quizzed.

"Yeah, we went into the shelter last night. All of the women and children have to go to the shelter."

"What about your father?" Moving forward, closer to the girl, the psychic tried to discover the strange circumstances behind the lost child's presence.

"My sister said he went to war." The tears finally erupted from the child's eyes as fright filled her. Forcing herself to speak, she continued to babble in fragmented gasps. "My sister said people die in war. I don't want my daddy to die." Whimpering, she trembled, her lip curled and tears still running down her face. "I don't want him to die so I came out to find him."

The news was unsettling. It appeared that the Slumlanders and the rat tribe were about to face off again. Everyone in the team knew that it would be a battle to the death. With the Slumlanders desperate for clean water and the reserves of the rat tribe dwindling from past conflict, everyone knew neither faction would stop the attack. If war erupted, one of the tribes would be completely annihilated.

"Come on, we might still have time to stop this mess!" Mineera yelled, propelled into motion. Charging once more into the fog, she disappeared, leaving the rest of her companions stunned.

With a sigh, Banion shook his head. "Let's go." Charging off, the tribals followed their leader, who was making an effort to keep up with the frantic Mineera.

There are some events in life which can change the fate of many in a matter of a few seconds. Mineera knew that this day was one of those days where every second was precious. With a resolute passion to stop further bloodshed from gripping the Concrete Barrens, Mineera sought to put an end to reckless aggression.

Her mind was spinning as she charged forward. In the past, she had used her skills to weaken and even destroy people with charmed words. Having spent a lifetime in the service of the Reaper Kai, Mineera had much blood on her hands and a stain upon her soul. Deep down, she believed in the good in all things, but her previous tactics in the service of evil had compromised thousands, leading to great suffering. Finally free from the darkness that had riddled her soul, the reformed demon knew that some way, somehow, she would make a difference. Failure in her crusade was not an option.

Letting the core of her soul lead her, Mineera charged through the ruins without concern for her own safety. The selfish part of her essence had been driven back. Instead, she had embraced an ideal of self sacrifice. If it came down to it, Mineera would make any sacrifice necessary to halt the aggression about to drive the region into acts of genocide.

A ray of sunlight had pierced the fog. Burning away the white darkness, it was a radiant spire of light charging down into the center of the city. Ahead of her, the Slumlander tribe was entrenched, holding defensive positions around a large open area. Across the wide, misty expanse, small beady eyes looked back at them. The rat tribe had taken a position across from their enemies, carefully preparing for battle.

Both factions had not yet drawn blood, but the tension was high. The fraught situation required only one rash act of hatred to plunge the region into a violent conflict. Both sides were hesitant to fight a final battle, a battle that one tribe would not survive. The stakes were high. Whoever the victors proved to be, they would be rewarded with a crown laid upon their heads, one forged with blood.

As she surveyed the forces dug in on both sides around the open courtyard, Mineera was undaunted. Witnessing the spike of light burning away the fog was a powerful omen. She knew deep down that it was a good sign. Understanding that stopping the conflict required a rash action, she took the initiative and charged forward into the courtyard.

The burst of movement was unnerving to both factions. Neither recognized her as an ally, but all knew her dark heritage. The traitor of the Reaper Kai jumped out across the Slumlander battle lines and charged toward the center of the courtyard. Amongst the ruins in the courtyard was a toppled pillar of concrete and it was her goal to reach it. Hundreds of weapons were trained upon her as she moved. Each faction was unsure of her intentions. With shaky, itchy trigger fingers surrounding her in every direction and a host of weapons targeting her, Mineera had put herself in a situation of mortal peril.

Her impulsive actions had turned both sides' attention away from the imminent battle, beguiling the would-be-combatants into watching her wild flurry of movement. With a leap, she climbed atop the fallen pillar and stood upright. Breathing deeply, she removed her hood and spun around, gazing directly at many of the warriors surrounding the courtyard. Making eye contact with as many as possible of both the ailing humans and the mutant rats was a powerful tool. Her bright blue eyes, shining out, caught many of the warriors off guard. Instead of focusing on the coming conflict, they gazed upon her graceful beauty. Holding her head upright, a sign of unwavering confidence, she was a powerful presence. The bold action had drawn the gaze of everyone about to partake in the horrid battle.

After securing their full attention, she inhaled, filling her lungs with air. A booming voice erupted from the dark skinned maiden, catching everyone by surprise.

"If blood is to be spilled this day, then mine will be the first to be shed!" Her commanding voice was chilling. Her assertion was so cryptic that many of the weapons on both sides were immediately lowered. Befuddled by the strange event, they all watched her curiously.

"If war is what you want, I offer myself as the first kill. I offer myself as a willing sacrifice!" She spun around, head held high, holding her hands outward in an act of submission. The crowd was stunned. Most were unable to comprehend the madness before them. But Mineera knew it wasn't madness which was driving her; she was being guided by an unwavering sense of righteousness.

"Get down from there!" the leader of the Slumlander tribe, Prefect Dale, shouted out, clearly enraged. "This is not your struggle! How dare you presume to know what has brought both of these tribes to this field of battle!"

Turning her attention to Prefect Dale, Mineera narrowed her eyes and locked them upon his own. The piercing gaze was aggressive enough to cause Dale's heart to skip a beat out of pure intimidation.

"This is my struggle. If needless bloodshed is what you are searching for, then you must kill me first," she said resolutely, defying the leader of the Slumlanders. "I will not leave this ground until the threat of war is abated. If it costs me my very life, if I need to sacrifice myself to show how futile this conflict is, then so be it. I give my life willingly to avert this tragedy. Let my death be a beacon to everyone in this city. You want needless bloodshed? I will be the first one to fall, a symbol of how futile war really is."

Mineera's powerful words were a potent image. The thought of attacking the defenseless woman out of anger seemed ridiculous, and made everyone in the two opposing armies think about their own conflict. Fighting and killing each other was beginning to seem foolish. Was war really worth the cost? Many were beginning to think that conflict was not the answer. Indecision was her intention, and she had succeeded in confusing many in both opposing armies. If they were confused, they could be made to listen to reason. As her act of selfless compassion tugged at the hearts and minds of both armies, more weapons dropped on each side, a sure sign that the seasoned diplomat had already made an impact.

With red eyes flashing, King Lavosi, the lord of the rat tribe, charged atop a pile of rubble opposite the Slumlander Prefect and taunted the diplomat. "Your pathetic act will not avert the inevitable!" His white fur standing on end, the enraged mutant screamed at Mineera. "Nothing can stop this conflict!"

"War is not an option at this point. There is no reason for this conflict to continue."

Shouting back at Lavosi only enraged him further. Quivering in wrath, and with no thought of his own personal safety, the hot-headed mutant leader was now standing in the open, fully exposed. In a time of war, this act would have meant certain death. But under the circumstances, it was irrelevant. Everyone in the crowd was too distracted by Mineera's bold actions to even consider killing him.

"You fool!" Prefect Dale boomed, coming out into the open as well, taunting her with his rage. "We are on the verge of death! We cannot yield or we face a slow extermination! Without clean water, we will die. Climb down from there and let us finish this war."

Mineera refused the order; instead, she stood with a smug look on her face. "The only fool here this day is you. During our absence from this city, we found a way to stop the aggression between your tribes."

"What are you talking about, witch?" Lavosi quivered in his place, his jagged teeth exposed in a sneer. The albino rat was a strange sight as he shook with rage, standing upon the rubble.

"I have a deal for both tribes!" she shouted, turning to look at both leaders in turn. "If the war ends right here, right now, I can offer a powerful gift to both tribes."

"Do not test my patience," Prefect Dale warned.

Mineera responded to his impertinence by staring him down. Finally surrendering to her formidable will, the leader of the Slumlanders fell silent.

"This is my deal. Both armies will leave this field of battle. King Lavosi will honor his deal with us, and give us the information we seek. In return, I will offer the Slumlander tribe operational water purifier parts. With a working water processor, there is no need for this battle to continue. With clean drinking water, peace can once again begin to grow," she explained, further stunning the crowd.

"Impossible!" Lavosi hissed. "No water purifier parts still exist in this region! You cannot stop a war with mere words!"

"Is it impossible?" Mineera quizzed the mighty vermin lord. "Tani, can you come up here?" The scholar was standing near the edge of the clearing with the rest of Nova 7, watching the intense situation unfold. With a gentle nod, he entered the open clearing. As he moved, everyone clambered to the clearing, completely disregarding their personal safety to watch the salvation of the region. Feeling thousands of eyes upon him, Tani sensed his face flush, turning bright red as he made his way towards Mineera. He reached the psychic standing atop the fallen column and handed her his backpack. She hoisted the satchel high and placed her hand inside.

Wild trepidation rolled through the crowd. With the threat of war still a tangible danger, everyone in the crowd secretly yearned for the claim to be true. If the satchel did contain the purifier parts, the war was indeed futile. As a collective gasp rolled through the crowd, her dark skinned hand emerged and held the electronic flow controller above her head. It was indeed true; Nova 7 had located replacement parts. A climactic release of tension rolled throughout both armies. It was as if the anxiety had been suddenly ripped away, leaving everyone breathing a sigh of relief.

"Impossible!" Lavosi hissed. Even though he was elated to avoid a battle, especially one in which he was severely outnumbered, and would almost certainly lose, the mutant was irritated that Nova 7 had chosen to disregard his orders to poison the Slumlanders. "So be it!" he screamed, red eyes bulging in anger. "You have a *deal*, foul Reaper Kai witch. I will *give* you the information you seek! But I warn you, do not test my patience any further!" With that, Lavosi turned away and gestured to his forces. The army of mutant rats followed and left the field of battle.

A sigh of relief washed over Banion as he saw one of the armies leave the courtyard. The situation was tense, and Mineera, by some wild act of selfless courage, had defused and ended the conflict which would otherwise have resulted in thousands of deaths. He stared in awe at the reckless Mineera, still standing like an angelic icon upon the fallen pillar. Such strength she had. He smiled and felt elated. The darkness that had gripped her, the insane voices, had been completely driven back. The psychic was free at last. Instead of a frayed woman driven by insanity, she was a powerful presence, someone who could change the fate of the world with her staunch will and charming words.

As Banion marveled at her courage and guile, Prefect Dale came before her and eagerly accepted the water purifier parts. For the Slumlanders, it was a glorious day. No longer did the threat of war hang over the tribe. No longer did they have to fight just to survive the harsh environment. No longer did the threat of being poisoned to death by toxic drinking water flood their minds. Instead, a ray of hope shone down upon the tribe. The Slumlanders could finally start rebuilding their world.

Chapter 25
Treacherous Deal

After the strange events of the day, the companions once again found themselves before the mighty doorway leading into the city of Verminhold. With the threat of war in the Concrete Barrens abating, Nova 7 had journeyed once more into the underworld to claim their prize, the information promised to them by King Lavosi. The goal of locating a viable nuclear warhead, which had been the focus of their quest for so long, seemed to suddenly hover within reach, and a nervous tension was gripping the companions' emotions. It had been a great many months since the fall of Rasheed, and finally, Nova 7 felt that they were about to take a giant step closer to ridding the world of the foul Reaper Kai.

Jared was feeling slightly rattled. Their surroundings were bathed in inky blackness, and the passage was strangely quiet. In the solitude of the moment, the tribal could feel the blood pumping through his veins and the sound of his own heart throbbing in his ears.

The young warrior touched the doors with his hand, brushing his fingers against the cold steel plating. Moving his hand downward, he searched for a way to breach the sinister obstruction. Intricate words in some foreign tongue were embossed upon the black metal surface. As his hand ventured further, the outline of the scribed words rose and fell under his fingertips, evoking a heavy sigh of anxiety.

"Is this a good idea, Banion?" Jared asked, wavering before the closed door. Still rattled over the day's events, he was unsure if

Lavosi would greet them as saviors or betrayers. The original plan had been for Nova 7 to help slaughter the Slumlanders. Instead, they had defied Lavosi's wishes and brought about an end to the hostilities in their own unexpected way. "There is a good chance Lavosi is going to be angry because of what we did."

"I don't give a damn what that vile rat thinks. The ultimate goal was to stop the Slumlanders from attacking the rat tribe. Instead of slaughtering the Slumlanders, we found an alternative to pointless carnage." Banion's voice was confident, and his eyes were shining aggressively; the other team members could tell that he was not going to take no for an answer. In Banion's mind, the rat king needed to pay them with the information that they were seeking. "He *owes* us!" Banion concluded in a harsh tone.

"I hope you're right," Tani responded with a pensive look upon his face. The tribal was staring at the mighty doors leading into the city of Verminhold with apprehension. "I hope he doesn't try to kill us."

"I felt his mind in the courtyard, when we confronted both the tribes. Lavosi hates what we have done, but he still saw the wisdom in our choice. Banion is correct; we stopped the threat, just in a different way," Mineera concluded.

"Who's to say that Lavosi even knows the location of a nuclear warhead? He could have lied to us," Jared said.

"I honestly believe the Runner knows the location of a nuclear warhead," Mineera countered.

"I hope you're right," Tani replied, sounding less than convinced.

"We saved his race, his entire city. We know it, he knows it. Even though we didn't fulfill our word as promised, we saved the Runner's ass, and its time for him to pay up. His tribe was severely outnumbered. The battle against the Slumlanders would have led to the annihilation of Lavosi's tribe. Even with all of his high powered weapons, he would have been overrun." Banion was harsh and assertive, getting straight to the point. "Open the door, let's get this over with."

The tribal's straining muscles rippled upon his arms. With a grunt, he pulled at the enormous door, which began to swing open. As the seal was broken, a whoosh of stale, moldy air erupted from behind the door. It took several tugs to move the creaking door sufficiently to allow the humans passage into the world of the mutants beyond.

A hiss and flash of movement erupted from within the dark passage beyond the doors. Hosts of rat warriors flooded from the doorway, pushing outward with great force. Jared was caught off guard and tumbled backward, landing on his rump.

The legion of rat warriors moved so quickly, it was futile to even attempt to grab a weapon. Banion simply held his hands in the air as an act of submission. The others in the team followed his lead. It was pointless to risk agitating the mutant tribe, especially after such a tense day.

The royal guards, Lavosi's personal aides and bodyguards, had been waiting behind the door. The leader of the guard was a stern, foul-tempered rat with brown stripes set atop a backdrop of black, bristly fur. Coming before the team, he hissed angrily, "I was wondering how long it would take to try and claim your prize, foul wretches."

"We saved your tribe, remember?" Banion shot back, none too keen on being bullied, especially by something which was only four feet tall.

"Saved us? You betrayed us!" another guard rasped, holding his assault rifle at the ready. Tottering forward, garbed in heavy body armor, the enraged mutant taunted Banion. "I should gun you down right where you stand!"

"Gun us down? If it weren't for us, you would have been overrun. I seem to remember that your tribe was severely outnumbered," Jared replied, face turning bright red, agitated by the brash actions of the mutant rats.

"Vile boy!" A booming voice rose from beyond the doorway. A white form was glowing palely in the shadows. A pink snout, blood red eyes, and sickly yellow teeth emerged from the dark passage leading into the city of Verminhold. Lavosi, king of dark places, approached, seeming none too happy. His pink albino tail flipped around violently, whipping the ground with a snap as Lavosi approached. Pacing back and forth as he snarled, he was definitely out of control. "What makes you think that you can come back here?"

"We had a deal." Banion took a step forward. The vermin bodyguards intercepted him with a hiss, but Banion was undaunted.

"Deal?" Lavosi screamed in rage. "We have no deal! You betrayed us!"

"Betrayed?" Mineera boomed. She moved forward forcefully, a presence to be reckoned with. "What do you know of betrayal?"

The comment was chilling and stunned everyone witnessing the exchange. The psychic had put herself into the fray by bringing attention to her own dark past. The statement could not even be questioned. Mineera had betrayed her own people, risking death to do so. Even Lavosi knew he had no authority to argue the question of betrayal in her presence. His rage was abated somewhat by Mineera's bold statement. Falling silent, he eyed the team while considering the situation. Staring at each of them, the vermin king fought an intense internal battle. After a moment's silence, Lavosi nodded his head as if reaching a decision. Changing the tone of his voice and demeanor drastically, he dismissed his personal guards. "Leave us!" he rasped.

Though stunned, they obeyed without question. Giving a final defiant hiss, the striped rat taunted Banion one more time before disappearing into the gates of Verminhold. With a grinding boom, the doors slammed shut, leaving Nova 7 alone with King Lavosi.

Banion's voice was still aggressive as he broke the silence. "You know damn well your tribe was outnumbered. Even with all of your weapons, the Slumlanders would have overrun and destroyed you. It's not a matter of pride; it's a matter of numbers."

Lavosi's simmering rage was beginning to abate. Deep down, the Runner knew that he could never hope to stand against the sheer numbers that the Slumlander army had brought to the battlefield. Banion was correct; it wasn't about pride, it was about sheer numbers. "What fate is this? Your quest is to crush the Reaper Kai with nuclear fire. You avoided genocide of one race to bring about the genocide of another race. You think you are all heroes? I think you are murderers! Murderers on a crusade of genocide!"

Genocide. In the end, that's what is was all about: killing an entire race. Nova 7 was on a quest to commit an act of genocide. Banion had refused to pull any punches with Lavosi, and the vermin lord had done the same with Banion's team.

Genocide. The word was still reverberating in each of their minds, a sickening, nasty word that evoked images of mass graves and swollen tombs. A reckless vision of shattered bodies, ripped and torn, filled their minds. The statement was chilling, especially coming from the vile creature before them. All of the glamour tangled up in the danger of their quest had been stripped away. All of the courageous heroism and the thoughts of justice had been torn free. All that remained was a horrible truth: Nova 7 was on a quest to commit mass murder.

"I would have slaughtered all of the Slumlanders and not had a second thought about doing so. Can you all offer the same pledge? Can you say that you could destroy an entire race and never have any misgivings? Is your passion to commit mass murder so strong that nothing can stop it?" the vile rat persisted in his taunts, feeling glee that he had put them all on the defensive.

Silence overtook the assembly. Lavosi's assertion was brash and aggressive. As they retreated momentarily, each member of Nova 7 felt a chill run down their spine. The reality of their goal had always been a distant thought, a cruel reality that they had all sought to avoid for as long as possible. Lavosi had brought the reality to light.

There were many times in their quest where they could have turned back. There were so many times that Nova 7 could have changed their minds and stopped the quest of madness. The team was tottering on the edge, a precipice where a single step could change their destinies forever.

It is at times like these in one's life that deep reflection is needed to ensure a solid course of action. Without a solid foundation, the road can grow dark and hostile. Without a solid foundation, destiny can be set in motion and nothing can stop the consequences of what has occurred. It was at this very precipice that Nova 7 rested, tottering on the edge of destiny.

Thinking about the horrible truth, they focused on the thought of mass murder.

"I never thought of it that way before," Tani said, subdued. At the beginning of their quest, they had been granted the status of heroes. Now the word 'murderer' was hanging over them. The young scholar had a hard time dealing with this transformation.

"But this is different," Jared countered.

"Is it?" the sinister creature shot back, red eyes flashing.

"Yeah, it's different. We are fighting for our survival."

"As were we!" Lavosi smiled as he spoke. "We have been fighting for our survival all these years. If not today, then some day soon, the Slumlanders will overwhelm us. How foolish you all are! You think because you stopped the war this day that it is done? The Slumlanders are now stronger! You didn't save our people, you delayed our death!"

"We stopped the threat," Banion boomed. "It's time for you to pay up."

"Bah! You still believe I owe you? Have you heard nothing? You delayed our eradication, nothing more!"

"You cannot know the Slumlander's intentions," Mineera shot back. "Allow peace to return."

"Peace?" he snarled. "Peace can only be achieved through war! Purging your enemies underfoot! Genocide is nothing more than a natural order, a natural response to hate."

"I am not a hate filled murderer!" Tani protested with a whimper.

"That is a matter of perspective. To your enemies, you are a loathsome thing, something that evokes the most passionate hate! To your allies, you are heroes, crusaders on a holy pilgrimage. The matter of genocide is a matter of perspective. But no matter how you sugar-coat mass murder, it is still murder!"

"I grow weary of such manipulation. The enemy cannot be bargained with or reasoned with. There is no such thing as diplomacy. I know that better than anyone. Our cause is justified since there is no alternative. There is no way to escape slaughter but to destroy the Reaper Kai order. The conflict is absolute. Only one victor will prevail. If we do not fight, we will die. If we do not resist, the armies of the devil will claim victory over the land," Mineera concluded boldly, staring at Lavosi with disdain. "If we do not make the hard choices, there will be no one left to resist the enemy. Such an act is unconscionable, but it is the only way. We either resist evil or we will be destroyed by it."

"There is no difference between your struggle and mine. The Slumlanders seek to destroy our race. We are no different, you and I." Lavosi pointed at Mineera. "Genocide is a matter of perspective. We are *all* murderers, only with differing points of view."

"I can never accept such a notion. Peace and the ability to live a calm existence are the true order of things. We are not murderers, just people who wish to live without the fear of tyranny. Justice is a force that brings about the absolution of truth, not violent slaughter. We do have differing perspectives, but reckless aggression in the name of natural order is a naïve assertion, one that should be relegated to fools." The words flowed from the holy woman with a crisp snap. Not the slightest bit of fear could be seen on her face as she confronted the loathsome rat king. "There is a profound difference between our struggle and yours. You would rather kill than negotiate. Peace can be achieved in your circumstance, but you choose to go to war. You lie to yourself and make yourself believe that murder is the only way. Your struggle has been tainted by a ruse. Since the foundation of your truths is a mystery, it's no wonder that you seek evil as the only way to satiate your twisted beliefs. We are different, that much is certain. We do not hide behind lies and half truths. The Reaper Kai order is ruthless, without a hint of mercy. We cannot negotiate and must fight. You choose to fight rather than negotiate."

Snarling, Lavosi shook his head and began to laugh at her. Staring at her with hate, he mocked her. "So foolish you are! You still will never understand. History will be the only judge of your actions. The victors of this struggle will either brand you as heroes or monsters, depending upon their perception. You still don't understand! Truth is based upon the victor of war, not reality. History is riddled with lies! Your struggle will be no different. Truth is forged by the strong, the ones that win and survive. The tainted and horrible deeds of those who win the war are forgotten. Mass murder is perverted in hindsight, changing from a horrible act of hate into a wondrous deed of courage. The perception of the strong dictates truth in this twisted world."

Lurching forward with a hiss, the rat king appeared to be on the verge of attacking Mineera. Seeing his advance, Banion came between them with anger flashing on his face. "Enough of your games!" the leader of Nova 7 yelled at Lavosi.

Smiling through his teeth, Lavosi looked at Banion. Their eyes locked and neither looked away, each answering the challenge. They stared at each other for many moments; all the while, the

Runner's smile broadened. A twisted glee was in his blood-red eyes. The creature was foul, cunning, and not to be underestimated. Flicking his tail once more, he spoke in an eerie tone. "Each of you must look me in the eye and acknowledge your intent. If you can do so, then you can have your prize."

Banion did not hesitate; his soul was foul and dark. Moving forward, he stared into the rat's crimson eyes, scowling at the mutant. Lavosi acknowledged him with a snicker. Banion's intent was clear: he had no misgivings about slaughtering the Reaper Kai.

The tribal warrior came forward next, his hand fidgeting with the raven totem around his neck. Touching the relic, a tremor rolled through him. Bracing himself, Jared looked into the sinister eyes of the rat king. Two blood red orbs stared back, their pupils hidden in the crimson flesh. Shuddering, Jared held his gaze with that of the mutant. He stared into the red eyes as long as he could, finally overwhelmed with a shiver. Looking away, the tribal felt he had crossed some twisted line, making a deal with the devil.

Mineera moved forward, undaunted by the vile creature. Staring into his eyes was not a challenge. She had spent a lifetime battling the will of evil. This creature was weak, in her opinion, a loathsome thing that deserved nothing more than pity. Satisfied with her conviction and strength, he nodded at her.

Tani was uncertain about facing such a challenge. The mutant was a terrifying thing to him, something that cared not for logic but only for bloodshed. Mustering his courage, Tani pressed his green eyes toward that fearsome sea of red. Those eyes were a lost place. To the tribal, the confrontation was nearly unbearable. Shaking, he averted his gaze quickly. Lavosi acknowledged him, tagging him as weak, a thing to be controlled and dominated.

The game had ended. The vile creature had assessed each of them, and now knew full well their strength or weakness. He smiled, knowing he had one more card to play from his stacked deck. "You will have your prize." He moved toward them with a sneer fixed upon his face, whispering, "The prize you seek is in the northlands, beyond the great forests, in a ruin so fearsomely protected, thirty of my own kind met a grisly fate there, inside the forgotten tunnels. I know now your resolve and it is formidable. So be it, fools. When the sun rises, *we* shall travel into the northlands and you shall have your prize."

An odd jolt of confusion rocked them. It took a moment for the sinister reality to bridge their synapses.

"What?" Banion said in a state of confusion. "Did you say '*we*'?"

"Of course. The road is long, you need a guide." The smirking mutant was elated by their response.

"Absolutely not!" Mineera was shocked at the prospect of traveling with the cruel Lavosi.

"I am not giving you an option. If you truly seek a nuclear warhead, I will *lead* you," Lavosi insisted, grinning wickedly.

"Lead us?" Tani whimpered.

"Yes, I will *lead* you." He spoke in a sibilant whisper, red eyes flashing with hate.

Shaking his head in disgust, Banion knew it was the only way. Lavosi would not simply give them the location; he would take them there, a dangerous prospect at best. Hiding his rage, Banion knew he had to deal with his adversary. Turning his gaze to the ruined tunnel, he spoke through gritted teeth. "So be it, Lavosi. You will lead us to the nuclear weapon."

In that horrible moment, they had made a deal, a deal with evil itself. Nova 7, once a group of four, was now a group of five. Feeling a knot form in their stomachs, they knew the road ahead was not going to be pleasant. Each of them looked at their new companion, who returned their gaze, smirking and staring at them with his crimson, hate-filled eyes.

Chapter 26
Dark Exodus

The ruined office building had become a place of great suffering. Strewn about the structure were the remains of four dead Slumlanders. Their mangled bodies had seen terrible torture, carried out at the hands of Reaper Kai priests. Having followed Guillotine's lead, the evil tracker teams had descended upon the ruins of the Concrete Barrens. Their intent was to locate evidence of a nuclear weapon, based on the conviction that Nova 7 would not be in the Concrete Barrens unless they had a credible lead to finding such a weapon.

Nova 7 had a five day lead, having already left the Concrete Barrens on their journey north to the military base housing the nuclear weapon. Only a day after their departure, eighteen Reaper Kai tracker teams intent on locating them had reached the Concrete Barrens. An enormous host of enemy agents was now bent on scouting the area for Nova 7's passing.

In an effort to find Banion O'Neil and his band of mercenaries, the Reaper Kai captured local inhabitants. Even after battering them beyond recognition, the sinister priests failed to unearth information about Nova 7. The torture of the Slumlanders had been useless. They were too primitive and feeble to even fend for themselves, and the Reaper Kai priests quickly discerned that the sickly humans were a dead end. The final Slumlander who had been tortured to death produced a juicy tidbit of information: he stated that the rat tribe in the area had an arsenal of Old World weapons. The Reaper Kai, already familiar with the legend of the Runner, immediately set out to capture members of the rat tribe.

The southern Darken Realm was rumored to be one of the best places in all the lands to locate ancient weapons. The Consortium of Arms, a legendary weapons trading guild, hailed from the Concrete Barrens. Everyone in the Darken Realm thought the guild consisted of human gun runners. In reality, the rat tribe, led by the Runner, was the real source of the Consortium of Arms. The Reaper Kai quickly pieced together the clues and immediately set out to locate members of the rat tribe to 'interrogate'.

Within hours, the priests had captured five mutant rat warriors and began to question them. As the priests were unperturbed by the constraints of morality, the questioning rapidly turned to violent torture. The deaths of the rat warriors revealed only that their leader, King Lavosi, was missing, and that his absence from the rat collective was a mystery.

From the torture of the mutant rats, the Reaper Kai had discovered that their king had disappeared on the same day in which Nova 7 departed. It didn't take a psychic to put two and two together; the Reaper Kai now knew Nova 7 was traveling with King Lavosi.

The Reaper Kai stepped up their operations, and within a mere few days, managed to flush out and capture one of Lavosi's personal bodyguards. Surpassing all the previous captives in his endurance, the guard held out for many hours, staunchly loyal to his own tribe and king. Finally, after more than sixteen hours of continual torture, the rat bodyguard divulged Nova 7's destination: a military installation far to the north. Using information from the bodyguard, the Reaper Kai began to craft a crude map and a set of instructions to reach the installation...

The screams of agony had finally stopped. The lifeless body of the rat bodyguard lay silent, still chained to a bloody chair. His death was a welcome alternative to the pain endured at the hands of the malicious priests. With flies already landing upon his battered body, the rat warrior was a testament to a hate-filled world where reckless ambition ruled the fate of many.

With a callous smile, the head Reaper Kai priest spat on the corpse.   In his hand were a crude map and a set of instructions leading to a nuclear weapon, the holy grail of artifacts in the Darken Realm.   The threat posed by the Reaper Kai order had now become much more sinister.   Old King Toil's original intent, back in Rasheed, had been to unite the free peoples of the land, and bring about the nuclear destruction of the enemy.   Instead, the enemy was now also hot on the trail of the ultimate weapon.   After the destruction of Rasheed, The Steel Crag Mining Guild, and the Mord Tech empire at the hands of the Reaper Kai, only the Iron Kai remained to handle the wicked onslaught.   Though powerful and stalwart, the Iron Kai could never survive the Reaper Kai acting as a nuclear power.

Satisfied with the map in his hand, the head priest glowered at the corpses before him. He spat one more time in the direction of the tortured victims, then unceremoniously left the horrific scene, letting the cold night air wash over him as he exited the building. As wicked thoughts rolled through his mind, he mumbled a dark prayer, paying homage to some loathsome demon bound to the spirit world.   At the conclusion of his demonic prayer, Brother Noxium opened his eyes and looked around for his most trusted aide.

"Where is Brother Donal?" the head priest hissed. Scanning the darkness, he saw a host of Biogtech soldiers surrounding the perimeter, their pasty white bodies glowing dully in the dim moonlight.   They rocked back and forth, emitting a small robotic chuckle every once and again.

A sudden quick movement, uncharacteristic of a Biogtech, caught the head priest's eye. Pushing forward from the perimeter was a squat priest with short brown hair and eyes wreathed with blackness. Even in the dim light, the Reaper Kai brother appeared to have two black eyes. In reality, his devotion to evil was so strong that the dark powers had bestowed upon him a grim visage, branding him as a true patron of evil. Brother Donal was a veteran priest with many miles and missions under his belt.   Breathing softly, he eyed the leader of the operation, Brother Noxium, with respect.

"How long have you been lurking about, skulking in the darkness?" Noxium hissed, staring at his trusted aide.

"Long enough to hear the screams." An abrasive, hate-filled look branded his features. Donal was twisted, truly twisted. Torture had always been a secret obsession for the dangerous priest. Though he didn't care to actually torture anyone, he was strangely fascinated with the screams and pleas of mercy erupting from the victims. This sick habit had become a routine, and Noxium knew that Donal usually hovered near enough to hear the sounds of torture, hiding somewhere in the darkness.

"I trust you have had your fill, then? If not, I am sure we could round up another victim from this cursed city. Have you had enough screams for one night?"

"Yes... now I am satisfied..." Donal whispered in a gentle tone, eyes shifting nervously around.

"Good, we have work to do. I have an assignment for you."

"What is it?"

"The dead mutant has given us the information we seek. We now have a credible lead on a nuclear warhead."

Donal's smile reflected his elation. With a nuclear warhead in the hands of the Reaper Kai, no army in the Darken Realm could oppose their will. "That is great news indeed!"

"The road into the northlands is long. Nova 7 has already eluded us once more, and we believe that they are traveling with the Runner, the damned king of the mutant vermin that infest the sewers of this city. I need you to take a few of the tracker teams and push on ahead of Nova 7."

"The road is truly long into the northlands. How am I to find them?"

"This map will be your guide." Noxium handed the tattered parchment to his aide. "I trust you can still commit anything you see to memory?"

"Indeed," Donal replied. Brother Donal's photographic memory made him an important asset to the Reaper Kai. The ability to sneak into enemy strongholds and steal complex documents simply by looking at them was a powerful skill, one that had earned Donal great status in the empire.

Concentrating, he stared intently at the parchment. The symbols and locations were burned into his mind. Within a matter of seconds, the process was complete; the sinister priest had memorized the entire map.

Noxium retrieved the parchment. "Close your eyes and listen to my voice," he commanded. "Can you see the place in the ocean called the Steel Loft?"

Donal concentrated, seeing the image in his mind's eye. "Yes, master."

"The coastline east of the Steel Loft is utterly inhospitable. In order for Nova 7 to reach the northlands, they must travel by sea. In order to keep from running out of supplies along the way, they must stop at the Steel Loft."

"I can see the place in my mind. What is your will?"

"Nova 7 has almost a week's head start. In order to assure our victory, I want you to take two tracker teams and set an ambush at the Steel Loft. Wipe them out, and our victory is inevitable."

"I will be hard pressed to reach that destination before them. We must travel night and day to reach the Steel Loft first."

"Indeed. You will set out immediately. Provisions and supplies have already been loaded onto a speedy ship. With the winds along the coastline, you should push quickly to your goal."

"Your will be done, master." Donal bowed, approaching the Biogtechs. He ordered two full squads to come with him, and the mindless robots followed without question. A female initiate of the Reaper Kai order stationed nearby was also ordered to go along with Donal.

And so it was that two Reaper Kai and sixteen Biogtechs assembled at a powerful clipper docked nearby. Boarding quickly, the agents of evil prepared to push northward, with the intent of ambushing Nova 7 at the Steel Loft.

With sails unfurled, the boat lunged forward and disappeared into the darkness of the night. Evil intent was the passengers' only companion. They had their orders and would do everything in their power to carry them out. Intent on executing the planned ambush, the sinister agents of evil moved with great speed to kill the members of Nova 7 and bring about the end of their crusade.

Chapter 27
Unlikely Saboteurs

A chill wind rose from the dark mountain peaks to the north. The frigid blast of air rushed downward across the jagged stone face of the mountain, plunging quickly onto the mountain pass. A lonely form huddled in the darkness of the night, sitting near a crude campfire to ward off the chill night. Alone beneath the enormous peaks of the Steel Crag mountains, the lonely wanderer was gripped with indecision, an unsettled irritation that failed to abate.

Looking upward into the darkness, the lonely figure eyed the snow fields reflecting the sparse light of the moon. Set amidst the dark backdrop of the shadowed gray rock, the snow field was in sharp contrast to both the night and the mountain to which it clung. As the lonely wanderer stared at the natural wonder, another blast of chill air rolled from the peaks with a whistle. Its intensity increasing, the gale howled across the jagged rock face above the mountain pass with a haunting wail. The chilling melody, rising and falling in pitch, made the barren landscape even more inhospitable.

The lonely figure shivered in the cold wind, then returned her attention once more to tending the campfire. Grabbing a clutch of branches, the young woman threw the fuel onto the fire. The flames grew in strength as they consumed the wood kindling, flaring into hues of bright orange. Popping and whining, the inferno ejected the cinders, throwing crimson ash into the swirling wind.

The crackle of the fire caught the attention of the mighty hippo Globulus, who emerged from the darkness with the intent of sitting down next to his tiny companion. Carla Reins was huddled near the camp fire, hands extended, warming her palms. As the hippo emerged from the shadowy night, Carla looked behind him as if expecting others to come slinking from the darkness.

"Anyone out there?" she asked in concern.

"Absolutely no one. Maybe we are in the wrong spot?" Globulus wondered, shaking his head in frustration.

"I don't think so. I grew up here in the Steel Crags and this is the spot. The other commando teams are just running a little late is all. They will show up in the next couple hours."

"They were supposed to meet us at sunset, and that was over two hours ago." Globulus shook his head once again in despair.

After the destruction of the Steel Crag Mining guild, the Reaper Kai took the rest of the region by force, subjugating all of the surrounding townships. The newly acquired territory was rich and ripe with resources. Once they had taken control of the region, the Reaper Kai gained access to nearly limitless supplies of both steel and petroleum, more than enough to create thousands of Biogtechs and other sinister war machines.

After the fall of the Steel Crag Guild, Emperor Gunther had ordered the destruction of the Gold Road, an enormous series of tunnels and mines beneath the Steel Crag mountains. His intent was to sabotage key tunnels within the mountain in order to prevent the movement of steel and other resources out of the region. Even though the enemy had an enormous cache of raw materials, with the tunnels collapsed, the Reaper Kai would be unable to effectively utilize them.

As a result of Gunther's orders, word was sent to the troops fleeing the mountains after the fall of Rust Spire, the Steel Crag Mining guild's former capital city. Many troops answered the call, and several commando teams were assembled to infiltrate the Reaper Kai-occupied tunnels of the Gold Road. The commando teams had specific orders and specific targets within the mountain stronghold. Using explosives to collapse key tunnels, they would attempt to cripple the Reaper Kai's ability to travel under the mountain and move resources.

Globulus and Carla were one of the commando teams that had volunteered to assault the Gold Road. The original plan had called for all of the commando teams to meet in the Steel Crags the night before the attack on the Gold Road. Unfortunately, Globulus and Carla were the only ones to reach the rally point. The other commando teams were absent from the rendezvous location, setting both Carla and Globulus on edge.

"No one else is coming, Globulus." Carla spoke in a dismal tone. She was still crouched near the fire, warming herself. In the flickering orange light, Globulus stared at his tiny female companion. He marveled at her delicate features. Petite and very deadly; that summed up Carla perfectly. She didn't appear dangerous... and this was exactly what made her so.

A moment passed, and Globulus conceded defeat as well. "I think you are right. The other teams must have never made it through the Reaper Kai defenses."

"What do we do now?" she asked, resting her head upon her knees, which were drawn tight against her chest. "Maybe we should just get the hell out of here? I'm not sure the two of us can make a difference at this point. We have limited explosives, and it will be difficult to break into the Gold Road undetected."

"It seems impossible, doesn't it? The two of us breaking into heavily patrolled tunnels..." Shaking his head, Globulus was also on the verge of giving up. Carla was right; how could the two of them make a difference? Wavering back and forth in his mind, Globulus waged a war with his own emotions. The easy road would be to run away, allow the Reaper Kai to move valuable resources, and thus enable the enemy to produce more troops. The tough road, the path wrought with peril, was to attack the Gold Road despite the danger.

"I hate the Reaper Kai," Carla exclaimed, almost laughing. Sighing with indecision, she stared at Globulus, looking into his beastly brown eyes, which were wreathed with worry.

"I hate them too. Let's bring the battle to them. I say we fulfill our pledge and attack the tunnels. Maybe we won't make it, but I'm sick and tired of running." Globulus felt his emotions overwhelm his senses. Fighting back tears, he brought himself back under control.

Carla saw his momentary surge of emotion and became concerned. "Are you all right?"

With a grumble of embarrassment, he shot back, "Yes! I'm fine."

"You don't seem fine." Moving closer to him, she tried to

discover the cause of his pain. "What's wrong?"

"Drop it. I'm fine. Let's focus on the mission." Pulling a map from his belongings, Globulus moved closer to his companion and spread out the parchment so both of them could view it. The rapid change in topic was disconcerting to young Carla, but she decided to let the matter drop, for now.

"Our original goal was the southern tunnels, near the train tracks. After looking at this map, I say we stick to our original targets. If we collapse these tunnels…" she pointed at several junctions under the mountain, "the enemy will not be able to move resources out of the Gold Road to the train tracks. Without using a train, the steel ore will be useless."

"Agreed. We'll stick to our original objectives. The conductor will meet us outside the tunnels after we blow the explosives, and we'll take the train out of the region."

"I only hope the enemy hasn't got to the conductor as well. Without his help, escaping the region will be nearly impossible."

"I agree, if the train engineer doesn't meet us, we'll be in deep trouble." Anxiety was flooding both of them. With a loud sigh, Globulus looked over at Carla. "Get some sleep. I'll take first watch tonight."

Laughing, Carla looked at Globulus, her eyes bright with apprehension. "I'm too damn nervous to sleep. I'll stay up with you for a while, watch the stars and listen to the night."

Nodding back, Globulus smiled at his companion. Even though he would never admit it to Carla, Globulus found her company comforting. A silence came between them as they looked at the stars. Lost in thought, each of them hid their fear deep inside their volatile minds; for when the sun rose, they would put themselves in harm's way, two soldiers against a multitude of enemy troops. Their mission was a dismal prospect at best, but it didn't matter. Somehow, they would both find the courage to confront their fears and triumph once more against an insatiably bloodthirsty enemy.

Chapter 28
Breaking the Supply Lines

In the darkness she waited, a patient hunter with only a single goal in mind. Her targets were strong and deadly, but it didn't matter. Where they were strong and reckless, she was reserved and patient, making Carla even more dangerous. Breathing softly, she whispered as her mind focused on a never-ending repetition of words, "I am a rock. I am a stone upon the earth."

The tunnels of the Gold Road had become infested as of late with Reaper Kai forces, laboring in the earth to harvest vital ore for their war effort. Having spent almost half an hour creeping up upon their position, the deadly sniper was a testament to a stubborn hatred. Alone in the southern section of the tunnels, young Carla knew she had to take out the Reaper Kai or their escape would be impossible. Globulus was much further back, planting explosives in the tunnel. The goal of the deadly duo was to hamper the harvesting of resources by destroying key junctions in the tunnels. Collapsing the tunnels with explosives would make collecting the ore nearly impossible. If their mission was a success, it would take many months for the Reaper Kai to dig out the tunnels and resume mining operations. Without resources, the Biogtech production in Detro Tech city would eventually grind to a halt, allowing the Iron Kai to eradicate the treacherous Reaper Kai.

But in order for their mission to be a success, Carla would have to kill the two Reaper Kai priests barring their escape. Without an open exit strategy after blowing the explosives, the battalion of nearly five hundred Biogtech soldiers protecting the inner tunnels would quickly make their escape impossible.

"I am a rock. I am a stone upon the earth. I am a rock. I am a stone upon the earth." Over and over, the phrases reiterated in her mind, making it impossible for the nearby psychics to sense her presence. Without concrete thoughts centered on an emotion, the psychics could not use their skills to sense her. Squinting with the rifle pressed against her shoulder, Carla aimed at the chest of a Reaper Kai priest.

Distress was evident upon his face. His powers were stretching outward, probing the area. From time to time, he thought he was being watched. Staring into the darkness, he saw nothing. Agitated by the feeling of foreboding, he walked frantically about, attempting to find an outlet for his nerves.

"I am a rock. I am a stone upon the earth. I am a rock. I am a stone upon the earth." Focusing only upon the phrase, the seasoned sniper did not allow any of her own thoughts to pass into her open mind. If young Carla thought about anything, the psychic would sense her presence, and Carla's life would be in danger. As long as she dedicated every thought in her mind to the phrase, the Reaper Kai priest was unable to locate her. The iteration was a stagnant thought, which did not allow him to latch on to anything concrete. If her focus and concentration were to fail, the powerful priests could use their psychic abilities to track her immediately.

As the psychic's red robes flailed about him, his erratic behavior began to upset his companion.

"Stay still, damn it!" the other priest hissed in annoyance. "You are making me nervous."

"I am making myself nervous. Something feels out of place, but it's elusive. I can't put my finger on it."

"Don't worry, we are hundreds of miles from the front lines. We are in no danger whatsoever."

Listening to his fellow priest, he felt suddenly foolish. Abandoning the thought that they were being watched, the priest let his defenses drop. "Ah, hell, maybe you're right." Sighing, he gave up his frantic pacing and turned to his companion.

The crosshairs of the scope finally came to rest. "I am a rock. I am a stone upon the earth." Holding in her breath, her trigger finger quivered. Bracing the sniper rifle against her shoulder, she exhaled slowly while squeezing the trigger. The hammer fell and the primer cap exploded in the barrel of the sniper rifle. The gun powder ignited, hurling the long wedge of lead down the spiral cut barrel.

The snap was crisp and alarming. The Reaper Kai priest took a perfectly placed round in the center of his chest. The deadly projectile slammed into his sternum, shattering his rib cage as the bullet ground through his chest. The blast was so powerful that the red-robed Reaper Kai was pulled off his feet and thrown several feet backwards. The hole blasted in his chest was a mortal wound which left him dead in a matter of seconds.

Stunned by the gunfire, the other priest looked in horror as crimson blood covered the floor of the tunnel. Only a few seconds ago, his companion had been talking animatedly. Now, in a stunning cascade of brutal reality, he was dead. Diving for cover, the alarmed priest hit the deck, moving behind a tall pile of ore. He hit the deck, and just in time. A second blast boomed near his ear and whizzed past his head as he dove for cover.

"Damn you," Carla mumbled, shouldering her weapon and going for her pistol instead. Without hesitation she rose, nearly invisible in her black attire. Holding the pistol in both hands, she crept forward silently as a cat on the prowl. Eyeing the rock pile in which the Reaper Kai had concealed himself, she prepared to ambush him once more.

"I need support!" he yelled excitedly into a radio hand set. "We are under attack!"

"Report your position," erupted the response from the radio.

"South tunnel! Hurry…" His speech was cut short. Rounding the rock pile, young Carla drew down on the Reaper Kai. Taking quick aim, she opened fire, shooting him four times with her pistol. All bullets bit flesh and the priest died instantly, slumping against the rocks.

The radio sprung to life in his dead hand. "Support on the way. Hold your ground, reinforcements are on the way."

Scowling, Carla's short, hooked nose trembled in anger. Holstering her weapon, her right hand brushed her short brown hair away from her ear, an unconscious nervous habit.

Running full speed, she hurled herself deeper into the tunnel, shouting urgently to summon her companion. "Globulus! We need to get out of here."

The mighty hippo warrior grimaced in anger, irritated by the ruckus she was making. He was fumbling with the explosives, having just finished wiring the last detonator.

Locating him, she rushed up to him, breathing quickly, her eyes darting back and forth. "I missed my second shot."

"I heard six shots total," he grumbled while standing up.

"First shot hit, next shot missed, last four hit. He got off a distress call on his radio."

"Damn it!" he boomed in disappointment.

A cackle rose from the tunnel. A host of Biogtech soldiers, over thirty in number, had shambled into the passage. Turning around, the mighty hippo warrior caught sight of them a split second before they opened fire. The thundering racket of submachine fire roared to life. Pulling Carla out of harm's way, he almost yanked her arm clean out of its socket. As he dragged her around a wall, the shower of gunfire rammed into the stone passage, ricocheting off the rock with an intense whistle. It sounded almost like popcorn popping, zinging back and forth as the bullets bounced around the underpass.

"Now what?" she whined.

"Run!" he boomed back. "We only have five minutes on those timers."

"Five minutes?"

Globulus was already on the move, and Carla followed quickly. "Yeah, you were supposed to kill them both, remember?"

"I did!"

"Not fast enough." Bolting down the passage, running toward the intersection, they were totally unprepared for what was about to happen.

Coming from the other tunnel toward the intersection was a small patrol of Biogtechs, two in number. The shambling robotic soldiers entered the intersection too quickly for Carla and Globulus to respond. Each rounded the corner too late. With a crash, Carla and Globulus ran smack into the sinister Biogtechs. Being far larger than any other participant in the collision, Globulus remained on his feet. The robotic death machines, in contrast, bounced off him and landed on their metallic rumps. Carla was caught in the mix and landed on top of the agents of evil.

There was a brief moment of chaos as each of them was

stunned by the encounter. Fortunately, the mighty hippo was the first to react. Without hesitating, he moved forward and roared out a fierce war cry. With a grunt, he brought the full force of his right leg down upon the chest of one of the Biogtechs. When he stomped with all his might, the strength of Globulus could not be averted. With a forceful fury, he crushed the robot to a pulp, compressing its chest like a grapefruit stuck under a steamroller.

Struggling to gain its bearings, the remaining Biogtech tried to defend itself but was too slow to escape the thundering force of the hippo. Grabbing the Biogtech off the ground with one fist crumpling its shoulder, the hippo warrior did not falter. His left hand began to pummel it without mercy. He struck the Biogtech repeatedly with his enormous, rock-hard fist, and the robot, stunned by the attack, was unable to respond. Energized by the conflict, Globulus punched the Biogtech several more times. With a look of disgust, he hurled its limp body against a nearby wall.

Another burst of gunfire rose from behind them. The war party had arrived and had opened fire once more. Several bullets struck the hippo upon his back. Barely breaking his tough hide, they caused minor damage.

"Move it!" he boomed, grabbing Carla off the ground. Bullets whizzed past them as they ran full speed down the tunnel.

The darkness of the passage was beginning to dissipate. Ahead of them, the tunnel ended, its exit flooded in sunlight. Black Rock Canyon and liberation were just ahead. "I hope your conductor friend made good on his promise."

The original plan had been to kill any guards, detonate the explosives, and escape by train using the help of Carla's long-time friend.

"He won't let us down," she replied.

The situation was now a little more frantic. Originally, they had not anticipated being pursued by enemy soldiers. Reaching the end of the tunnel, they rushed into the warmth of the sun.

Ahead of them was a train with only a few cars. The locomotive in front was belching a trail of black coal smoke, snaking into the bright blue sky.

Seeing them emerge, the aged conductor smiled and waved at them. His smile quickly turned to alarm as they frantically shouted at him. "Get it moving *now*!"

The conductor's concerned face quickly became bright red, as the urgency of the situation registered. Throwing several scoops of coal into the engine, he slammed down the plate and the fire erupted in the engine. He pulled a lever, and the engine clutched the wheels, which turned over slowly. The engine lurched forward, inch by inch, momentum slowly building.

"You have got to be kidding me!" Globulus boomed, seeing how slowly the locomotive was churning to life.

"Now what?" Carla yelled in alarm.

"Jump into the ore car."

They both launched themselves into the steel-lined ore car. Emerging from the tunnel, over fifty Biogtechs were about to catch up to Carla and Globulus. Though slow, the horde of Biogtechs was currently quicker than the lurching locomotive.

Bringing his brace of guns from his belt, the mighty hippo palmed the pair of shotguns like a human would grip a pair of pistols. Well within range, he opened fire, sending a spray of pellets into the host of robotic soldiers. The scatter gun was an effective choice against the massed troops. The shots hammered into them, spraying hydraulic oil and blasting plastic parts free.

"Don't just sit there! Help me!" he boomed just as the first wave of return fire hurled forward. The metal-plated ore car was a great defensive position. The submachine gunfire ricocheted off its walls harmlessly.

Brandishing her pistol, Carla looked over the edge of the car, and fired several well-placed rounds, felling two robotic soldiers.

The Biogtechs had collected their bearings and eyed the duo with disdain. Over forty trained their guns upon them and opened fire.

"Hit the deck!" Globulus screamed. Carla, seeing them about to fire, had already crouched down. Hundreds of bullets hummed forward, spraying the ore car with a shower of lead. Most were turned aside by the steel plating, but the intensity of the attack had weakened the plating in several spots. The failing plating was never meant to withstand gunfire, and several bullets pierced the steel.

Hearing the bullets whiz past them while inside the steel car was unnerving.

The conductor was frantic, sweat pouring down his face.

Shoveling for all of their lives, the aged man hurled piles of coal into the engine. The fire burning inside was intense, driving a fresh wash of power into the locomotive. Black fumes belched from the stacks as the train began to pick up speed.

Clunk, clunk, clunk, the wheels rolled over. The Biogtechs continued the assault. The steel plating was now riddled with holes in several places and the bullets were ricocheting inside the car. Shielding her head with her hands, Carla closed her eyes and listened to the fierce attack.

Globulus, positioning himself between the attackers and Carla, had sustained several gun shot wounds. His thick coppery brown hide was bleeding in several places as he tried to shield his companion.

The engine was finally primed. With the wheels of the locomotive chugging along at a quick pace, the train moved down the tracks, outpacing the slow rabble of Biogtechs. A moment passed, and the train outdistanced the effective range of the submachine gun fire.

With a sigh of relief, Globulus looked at Carla, still crouching in a defensive position. "It's all right, lass, its over."

Looking up at him, she gave a sigh of relief. As she played nervously with her short brown hair, the adrenaline of the conflict began to abate.

"Next time, kill them quickly!" the hippo roared comically.

Smiling back, she rolled her eyes. "Next time, don't take so damn long planting the explosives."

The train rolled down the tracks, moving eastward down Black Rock Canyon. In the distance, a dull boom thundered. The demolition charges set by Globulus had detonated underground. With the fire from the explosives came ruin. Several tunnels, rocked by the shockwave and energy, collapsed under hundreds of tons of falling rock. The mission was a success. The courageous young Carla and Globulus had destroyed the mining tunnels and the enemy's ability to produce viable resources. It was a small victory, but a victory nonetheless.

Chapter 29
The End of an Age

The train was steadily pushing east down the canyon. The battle-worn companions, fresh from sabotaging enemy mining operations, were silent. The events of the day had worn them both to a frazzle. Though still stunned from the reality of their escape, Globulus and Carla were slowly coming out of their daze. The burst of adrenaline was gone. All that remained was a sullen, empty feeling, a feeling somewhat akin to coming down from an alcoholic binge. During a wash of alcohol, one could feel wonderfully dangerous. But after the high, all that remained was a drowsy, sleepy sickness, cloyed with depression, leaving no option but to hold on until the nausea subsides. Such was the sickness of combat, a rush that dwindled to an oppressive memory after the fact.

No matter how many lows followed the rush of combat, there was something euphoric about each and every conflict, bringing the adrenaline junky back to that dangerous precipice where life and death seemed to blur together. The highs and lows would begin to wear away a person's most staunch defenses, eventually making them crumble completely, searching for other avenues of escape.

The stupor was abandoning both Globulus and Carla, leaving them wide eyed but exhausted from the ordeal.

With dull fascination, they looked out of the ore car, viewing their surroundings. With each chug and churn of the locomotive's engine, the train pushed deeper down the canyon, a place that had been active with life only a few weeks prior. The once-thriving towns had become strangely quiet and devoid of life. Passing by the ruins, their somber gazes could not help but linger.

The recently occupied remnants of the Steel Crag Mining Guild were now under the control of the Reaper Kai. With Rust Spire the center of enemy troop presence in the region, other, smaller towns and provinces belonging to the once-mighty guild had all been abandoned. The threat of attack had driven everyone away, leaving dozens of ghost towns throughout the area, especially in the canyon.

The silent husks of houses and shops lay dormant. They were once places teeming with life, places where the hopes and dreams of the towns' inhabitants could be realized. The people of Black Rock Canyon were a peaceful people. Now they had moved on to more civilized lands, beyond the drums of war and the hunger of conflict. The silent towns were a melancholy, oppressive scene to witness.

Flourishing gardens with flowers were still in evidence, blooming in the warm sun. Garden tools had been unceremoniously left near the scene, another silent reminder of the urgent evacuation. The once-bustling avenues were devoid of life, an eerie testament to fear. The only thing moving amongst the forgotten byways were dust devils, prancing about in circular motions as the wind rushed down the open streets. Though the town itself was lonely, it was the abandoned houses that gave Carla and Globulus a true sense of forlorn loss.

White curtains, lovingly cleaned, hung in the windows of the houses. Bright colored paint had been carefully drawn across the exterior of the homes, giving each one a unique feel. These were not just homes, but a part of someone's dreams, part of a life now gone.

It was eerie to behold such a quaint town, abandoned to the reckless silence.

As the train passed ghost town after ghost town, both Carla and Globulus knew it was the end of an era. Both knew that an age of peace and prosperity had come to an end. The armies of evil had taken control of the region, and it would be a long time before the tide of darkness could be driven back.

The conductor was silent, his expression grim. He was unable to speak as they passed through the graveyards of civilization. Carla, on the other hand, had emerged from the dreary depression of her combat high and was staring at her beastly companion. Blinking several times, she was suddenly alarmed. The jagged wounds, bullet holes torn in the hippo's flesh, were now clumped masses of clotted blood. Her heart lurching, she moved forward to tend to his wounds. Bringing forth some bandages from her pack, the young woman began to treat him.

As she worked diligently, he shifted in discomfort, back stinging in pain.

"Hold still!" Carla whined as she cleared the dried blood from the wounds on his back. She was prying bullets from his thick skin using a blade. His coppery brown skin was thicker than the toughest leather. Making slow but steady progress, Carla had already managed to liberate three bullets from Globulus' back.

"You could be a little gentler," the mighty hippo said gruffly.

"You shouldn't get shot so often," she replied with a giggle, trying to distract the hippo with a barb.

"Keep it up, brat, keep it up."

"Now seriously, I wouldn't have to pull these bullets out of you if you were a little quicker."

"Me? Be quicker? Who was it again that *failed* to kill a few measly Reaper Kai? Was that me? I don't think so!" he grumbled.

"Oh no, you're not gonna bring *that* up again, are you? It's not really fair."

"Fair? Don't whine about pulling bullets out of me when it was *you* that alerted the whole damn mine that we were there. I am surprised we made it out at all."

"Fine!" Carla shot back with some venom in her tone.

"Is that attitude you're giving me?"

"Yes, yes it is. Can we just drop it? I was just poking some fun at you." With somber eyes and a pout on her lips, she blinked several times at him, almost looking as if she was about to cry. Her lower lip quivering, she averted her gaze. Looking back at her, the hippo's brow furrowed. Feeling a bit ashamed, he reached out his massive arm to console her.

"I am sorry…" he said with earnest concern.

With a beaming smile, her somber eyes brightened and a wicked grin crossed her lips. "Gotcha!"

He looked away with a scowl, feeling foolish. Shaking his

head in disbelief, he wondered how he could have fallen into one of her traps again. "Can't we just sit here? Do we have to babble another afternoon away?"

"Yes, we do. What else do we have to do?" Moving behind him, she finished binding his wounds. With a chipper tone, her fast paced, auctioneer-like voice began to rattle on about anything and everything. "This is great, being out here again. Before you met me, you had never been on a train. Remember that? Now in just a few months, you have two train rides under your belt and both on the same train! Imagine that. Don't you find that interesting?"

"No."

"What about the conductor? Do you think he's upset?"

Shaking his head, Globulus knew escaping the conversation would be impossible. With a sigh, he prepared to engage in some sort of never-ending exchange about the conductor's *feelings* and his *emotional state*. "No, he is not upset."

"How can you say that? Have you looked at him?" Carla spoke in a whisper, looking up at the engine compartment.

"Yes, I have looked at him. He seems fine."

"How could he be fine? His whole way of life is over. His work, his true passion in life, is gone." Raising his brows, he simply stared at the chattering nightmare. Undaunted by his silence, she continued on in her boisterous, shrill voice. Listening to her talk was sometimes like listening to someone screech their fingernails on a chalk board. "Look at him. He has been conducting this train his whole life. Now he has nothing."

"I don't care," he grumbled, feeling like covering his ears.

"It must be sad, working his whole life on this train. This train is probably the only home he has ever known. Think about having to give that up. With the enemy occupying this territory, this is his last ride. His train will never come back down these rails. Imagine what that must be like?"

Sighing again, the hippo did begin to think about the aged conductor, almost against his will. Having spent an entire lifetime on the rails, the loss of his daily routine would be a hard thing to deal with. Globulus was filled with a sudden empathy, and a keen sense of sorrow washed over him. Though he was a creature of war, born to fight, even the mighty hippo yearned for a simpler time where all living things could exist in peace. It brought sadness to him, thinking that this man's way of life was over. The one thing that he loved most in life, traveling the rails, was at an end. This ride would be the last one he would ever make. He would surely miss the smell of coal smoke wafting from the stacks and feeling the train lurch forward, and would undoubtedly never forget the sensation of listening to the engine chug away, and the anticipation of passing through each town. The smiling faces of all the travelers would now pass into memory. War is a state of strife accompanied by many costs. Even the life of the simplest of men can lose its charm when the gears of devastation collide with civility.

Passing through the ghost towns only reinforced this idea. How many people had lost all of their dreams? How many people had lost the simple joys in life? It was a staggering thought, washing over them as they viewed the abandoned homes and shops. In that moment, the reality hit both Carla and Globulus.

The shops weren't just abandoned buildings; they were abandoned dreams. The ghost towns were not just a symbol of lost possessions, but garish reminders of a lost way of life and lost ambition. Each building was a reminder that an individual had lost their passion in life. Each town was a reminder that an entire community had lost the same. How many lost dreams had they passed? How many abandoned hopes of peace had they passed on their journey down the canyon?

With a nod of compassion, Globulus acknowledged that he did identify with the conductor. "Aye, lass, I feel bad for him indeed," he said gravely, turning to Carla. "It must be hard to give up the only thing you have ever known, especially when you love it so much."

"I wonder what he will do now? What road will he walk?"

"There are many possibilities. I only hope he can find peace in this dark world."

"Me too." Carla smiled. Her short brown hair was blowing in the breeze. With a distant stare, she climbed up the side of the ore car. Bracing her arms over the rim, she looked toward the front of the train.

The conductor was shining the gauges on the engine with a rag. He smiled, his hand moving as it always had, cleaning the panels and making everything look perfect. He knew deep down that his days on the rail were over. Even though it was the last journey of the mighty train, it didn't matter. This was his baby, his pride and joy. The engine would look just as beautiful on its final voyage as it did on its first.

Staring forward, they discerned that the final leg of the journey was at hand. The train had passed out of Black Rock Canyon and drawn closer to the desert sands. The small train station of Hundred Palms was the end of the road, the last outpost on the rail line. Looking away, the conductor fought the sadness within him. His hand shaking slightly, he sighed, gripped the brake, and pushed it forward. The train began to slow. The engine began to churn with less intensity. The steel beast chugged softly as the final puffs of smoke rose from the stack. Finally, the train came to a halt at the abandoned station.

Placing his hand on the engine, which was now growing cold, the conductor closed his eyes for a brief moment. With a pat, he said goodbye to his train. His voice crisp, he boomed out, "Final stop! Hundred Palms!"

Carla stared at Globulus and he looked back with a look of compassion. Dismounting the ore car, they grabbed their gear and walked toward the front of the train. The conductor took off his hat, his expression somber.

"Thanks for your help in this mess," the mighty hippo boomed. "If it were not for your courage and commitment to aiding us, escape would have been impossible."

"It was my pleasure." He smiled, eyes darting back towards the train. "After the fall of the Steel Crag Guild, there was no use for me and my train. I am glad we were able to give it one last pass down the rail."

"We are too." Carla smiled warmly, grabbing his shoulder and rubbing it lightly.

"We are headed northwards, back to Iron Kai territory. You can come with us," Globulus offered.

With a sigh, the conductor shook his head. "I have a few things I need to pick up. I'm gonna head east, I think. Family lives close. I want to spend a little time with them before things get much worse."

Carla looked at Globulus with empathy in her eyes, and the mighty hippo returned her gaze, nodding in agreement. For a brief moment, both Carla and Globulus were on the same page and shared an unspoken idea. Carla was the one to turn to the conductor and voice this idea. "We can take you there, escort you to wherever you need to go."

A moment of uneasy silence followed. The conductor was staring at them blankly. Finally he spoke. "I need to be alone. Thanks for the offer but I need to be alone."

"Fair enough." Globulus bowed before him. "Thanks again for your courage. You aided not only us, but the war against the Reaper Kai."

"Farewell." Carla grabbed him in a heartfelt hug.

"Farewell, young Carla and Globulus."

Waving goodbye, the duo pressed on, moving out of Hundred Palms toward the northeast, down a wagon-rutted road.

Once he was left alone, the conductor returned to his train. Placing his hand on its side, he said goodbye one last time. "I will be back, mark my words," he whispered. "I will be back." With a smile, he turned away and grabbed his gear. Leaving the ghost town behind him, he headed east into open desert.

In the silence of Hundred Palms, the engine lay dormant, its coal fired engine growing cold. With war waging everywhere in the Darken Realm, its silence marked the end of an age, and the dawning of an era where peaceful dreams were becoming mere whispers.

Chapter 30
Of Valor and Cowardice

Lush green plants erupted along the white alabaster wall, pushing upwards. The shimmering white marble, in stark contrast to the bright vegetation, was an arresting sight. The interplay between the brilliant green and vibrant white hues made the lush garden look like something out of a fairy tale.

Violet flowers popped up at regular intervals, clinging lightly to dried sprigs and vines. Over the span of many years, the plants' tendrils had found small depressions and cracks between the marble stones, allowing them to slowly ascend the mighty walls. Gripping the niches, the vines and flowers rose several stories along the wall of the circular tower.

Birds, exotic bright blue songbirds, chirped and chattered, sitting high atop the well-tended shrubs in the garden. Bright red cardinals sang intricate songs, warbling out a series of musical notes that would have made any musician envious. Amongst the peaceful flowers and lush gardens played two children, born of different worlds, driven together by fate.

"You can't catch me!" the young hippo yelled in glee. Rushing through the garden of the Rasheed palace, young Globulus was smiling, his elongated tusk-like teeth erupting from his mouth. With a wide grin upon his face, the hippo youngling was enjoying himself greatly.

A girl, a few years younger, with pristine features and flaring blue eyes, rushed after him. Joyously laughing, she reached out, trying to grab the young hippo.

More stocky and slower, young Globulus was no match for the princess of Rasheed. Bounding forward, she extended her hand and grabbed his arm. The game had changed for the time being. "You're it!" Marion laughed. With a quick move, the young princess bolted off, away from Globulus, after 'tagging' him.

"I'm gonna get you!" he boomed back. With a determined look, his dark, beastly eyes fell upon the child princess. Mustering all of his speed, he charged forward like a crazed bull. As he lumbered toward her at full speed, she was shocked by his aggressiveness. A smile graced her lips and she stood her ground. Watching him carefully, she timed his advance. When he was but a few footsteps away, Marion jumped to the side and he crashed forward. Attempting to react to her maneuver, his right foot went one way as his left foot went the other. He lost his balance, crashing to the ground, his knees digging into the lush green grass.

"Ack!" Globulus yelped as his momentum carried him forward. Unable to stop himself, his face hit the ground, chin digging up even more grass and earth. As he lay face-down upon the ground, Marion came forward, laughing and giggling. Spitting some of the turf from his mouth, he rolled over and grabbed her leg. She fell forward, landing on top of him. Both of them began to snicker. Grappling with one another, the two children wrestled.

"Marion!" a piercing voice ripped the air, so chilling in tone that the songbirds were startled and flew away with great haste.

Ignoring the piercing voice, the two children continued to grapple.

"Marion!" The voice erupted once more, this time in a chilling rasp that seemed to freeze the air with an icy blast. The enraged newcomer was not to be messed with or disobeyed.

Responding with alacrity, the two children knew whose voice it was. The hateful queen of Rasheed, Asagara Toil, was bellowing at the two youths. There was a flash of madness in her eyes as she surveyed them. With a jolt of panic, young Marion Toil and Globulus jumped to their feet and returned her gaze with tense stares.

"So this is where you have been all afternoon!" The woman, beautiful in appearance, surveyed the children with a fearful gaze. Her blond hair trembled as her form was wracked with a simmering rage. Wringing her hands together, she sneered at her daughter.

Taking several steps back, both of the children watched her very carefully. The queen of Rasheed, Asagara Toil, was not known for her patience.

"We were…" Globulus offered meekly, trying to explain the frivolity.

"Silence, you sickly, fetid creature!" Asagara boomed. Her gaze fell upon the young hippo.

Globulus stared back with fear in his young eyes. Fidgeting, he backed away from her. The simmering rage coming off her was so intense, young Globulus could feel it.

Asagara returned her gaze to her own child.

"You have not finished your studies, Marion. Get back to the library and finish your readings. A foolish princess is a weak princess. Do you want to be weak?"

"No…" she whispered.

"No?" Asagara moved forward and grabbed her daughter by the wrist. For a split second, Marion thought her mother was about to strike her. Flinching back, the child expected the worst. Asagara's demeanor changed immediately as she saw her reaction. Crouching down, she looked into her daughter's bright blue eyes. Young Marion stared back and a woozy feeling washed over her. With eyes locked, the two stared into each other's souls. Smiling, Marion's sinister mother took her hand and brushed aside the blond tendrils of hair. Breathing in quick, hurried gasps, young Marion looked at Asagara very carefully.

"If one is to rule an empire, one must have fierce convictions. Without the study of one's heritage, one cannot know their place in this world. I want you to always know your place in this world. The weak, pathetic whelps that foolishly waste the day away will find the world has passed them by. Do you understand what I am saying to you?" Asagara quizzed her child.

"Yes, mommy," she said in a somber tone.

"Get back to your studies. If you get done with all of them before dinner, I have a *special* book to show you." Cupping her daughter's chin in the palm of her hand, she squeezed it tightly, staring into the child's eyes…

Globulus shook his head in dismay as the vision of his

childhood passed out of his mind. "I don't want to talk about it anymore. It's too painful."

Carla was silent for a brief moment, but then continued to pressure him anyway. The duo had been traveling together for months and it was the first time he had ever mentioned his youth and Marion Toil in such detail. The topic of the dark queen of Rasheed was a sore one for the mighty hippo warrior.

Eyeing him carefully, Carla knew that he had been hiding his emotions since the traitorous Marion Toil had ordered the assassination her own father and the destruction of Rasheed at the hands of the Reaper Kai. Though she knew he was uncomfortable with the topic, her own curiosity got the better of her. "You didn't do anything wrong."

"I didn't say I did anything wrong," the mighty hippo barked back at her.

"I know you didn't say it, but that's what you are thinking."

"Bah! Leave it be. You don't know anything about it."

"Don't I? I have heard the sorrow in your heart many times as you speak about King Toil. In all respects, he raised you. Don't beat yourself up."

Globulus did not respond; instead, he continued down the worn trail. The afternoon sun had passed beyond a ridge of rocks to the west of them. The duo was now traveling in the shadows, enjoying the refuge from the burning heat of the harsh desert sun. Getting lost in his surroundings, he tried to ignore the raw emotion in his heart.

"It's not your fault," she said softly, walking beside him down the path used for hundreds of years by merchants and travelers.

"What's not my fault?" He shook his head, grinding his teeth in frustration. His irritation at her intrusions was mounting, but his soul was also tortured by the horrific events which had befallen him over the past months. In a short span of time, the mighty hippo had lost his foster father in Rasheed and his real father after a heartfelt reunion. Instead of coming to terms with the loss, he had kept it to himself, internalizing the trauma.

"The death of King Toil and your father is not your fault."

Like a wild tiger, he bared his teeth and roared at her. "Keep your mouth shut!"

"It was not your fault," Carla persisted.

The duo paused in their path as Globulus averted his gaze. Shaking his head back and forth, he wanted to flee from her. His heart pounded in his chest as a knot formed in his stomach.

"It was not your fault," she pressed again.

"It was my fault. It was my place to protect the royal family. All the while, that scheming Marion was plotting the demise of the entire empire. She was right under my nose the whole time, and I lacked the wisdom to see her for what she really was. It was my fault that King Toil died. I failed in my role as his protector and it cost him his life and the lives of all the innocents in Rasheed."

"You didn't seek out to murder the king – his own daughter did. Your goal was to keep the entire royal family safe, including Marion. There was no way for you to know of her betrayal."

"But that's just it: I did suspect her. Before the fall of Rasheed, I saw her consorting in secret with an assassin. Instead of doing what was right, I failed to act. I let my own sense of duty to Marion cloud my judgment."

"And why wouldn't you trust her? You grew up with her and took a vow to protect Marion. Your sense of duty is not a curse. You only tried to do the right thing."

"But in doing so, the most courageous leader Rasheed has ever seen was put to death. I have much blood on my hands. I am a coward and deserve nothing." Globulus moaned in pain and exasperation.

"Coward? What kind of talk is that? A coward does not put his own safety at risk to protect the innocent. You are the most courageous, selfless friend I have ever had. If you cannot see your own strength and valor, you are a fool, but you are no coward."

"You speak of valor. But where was my valor the night Rasheed was laid to waste? Where was my courage when my father stood against an army of enemy soldiers alone? It took me a lifetime to gather up the courage to find my father. My cowardice allowed me to live when I should have died beside him, in battle, where it mattered most. I have lived twice where a hero should have perished. Do not speak of courage and valor. Where it mattered most, I fled the field of battle to walk the hollow path of cowardice." The statement was a wrenching one. Deep down in his heart, Globulus felt that he had failed the two most important people in his life.   His foster father, King Toil, and his real father, Chief Stoneskin, were dead.   In his mind, Globulus would rather have given his life than live on in the shattered memory of their former glory.   Shaking his head in sadness, he looked at Carla Reins and spoke in a whisper.   "I failed them…"

Carla felt the tears well up in her dark brown eyes.   She held back the flood by blinking several times.   Grasping his arm with both of her hands, Carla consoled him as best she could.   "As long as you keep them in your heart, you will never fail them."

"I miss them so much…"   A stream of tears flowed down his cheeks.   Shaking his head in frustration, Globulus became tense. The tension rose in him like a storm building inside his heart. Thinking about King Toil and Chief Stoneskin, he let the thunder flowing through him rekindle his desiccated heart.   The thunder built and mixed with his shallow rage.   Flexing his hands, a feeling of empowerment flooded him.   In that moment, he was not a coward wracked with self pity; instead, his heart pounded and he felt truly alive.   Eyeing Carla, Globulus made a vow.   "Mark my words here, this day, young Carla.   When next we meet our foes on the field of battle, I will not flee, I will not waver, and I will not let anyone dissuade me from a course of violent action.   I have lost much and wallowed in self pity.   I can do so no longer.   I would rather die than live my life as a coward."

The oath was powerful, and young Carla felt her heart lift. Where once there was sadness, a staunch power of will had now formed inside Globulus' once-fragile mind.   No longer was he a victim; instead, he was something much stronger, much more menacing.   Carla marveled at his strength and felt emotion overwhelm her.   A tear rolled down her cheek as she looked at him.

Smiling up at him, she wiped the tears from her face.   "You feel better now?"

With a sheepish look, he nodded his head in agreement. "Yes."

"Good, cause I am getting tired of standing out here in the middle of nowhere!"   She prodded him.   "The longer we stand here, the more you are beginning to smell."

"Come here, you brat!" he boomed.

"Gotta catch me first."   Carla smiled back.

He eyed her with determination.   As he focused on catching

her, he grew stern and serious. With a burst of speed, he thundered toward her. His bulky frame hurled toward her with frightening speed. Unnerved by his sudden burst of energy, she watched him lumber forward. With a determined look, she tried to time his advance and dodge his charge, sidestepping him with an agile move. But the momentum of his charge could not be abated. As he twisted toward her, she realized she was too slow. Grabbing her with his enormous hand, he lifted her off the ground and slung her over his shoulder.

Hanging upside down, she flailed her legs around, squawking like a captured bird. "OK!" she whined. "You win!"

Globulus put her back down on the ground with a smile. Giggling, she rearranged her disheveled gear and equipment. With that, the two friends took a step down the worn pathway, pushing slowly onward to the northeast, toward the refuge of the Iron Kai empire.

Chapter 31
Swath of Destruction

Towering well over twelve feet high, the enormous cactus stood defiantly against the harsh climate. Several stalks rose outward from the trunk, with a host of long spines jutting from each appendage. Bright red and purple flowers crowned each arm of the mighty cactus. Desert birds, brown in color, were perched atop each stalk. As the companions approached, the birds squawked and started flapping their wings, finally retreating with angry screeches.

Taking a brief break from walking in the harsh sun, Carla wiped away the sweat rolling down the side of her chin. Breathing heavily, she grabbed her canteen. The water was warm but was wet nonetheless. She tipped the canteen back, letting the water flow into her mouth. The moisture felt refreshing as she swished the liquid around. Her dry, cracked lips were momentarily relieved.

Capping the canteen once more, she looked upward at the massive cactus growing out of the center of the pathway which cut through the wasteland. Smiling, she reached out to pluck a flower off the monstrous plant, but as she extended her hand, she was stunned by a prick of pain. A small spine had pierced her skin. Reeling back, she watched as a tiny drop of blood formed at the tip of her finger. She shook her head in dismay, licked the wound and rubbed the left-over blood on her black military fatigues.

"You coming or what?" she inquired harshly. Looking back down the trail, she saw that the mighty hippo was winded. He was doubled over a fair distance away, breathing heavily. Seeing Carla, he felt foolish and began to walk quickly toward her. The massive hippo refused to concede defeat, especially to such a dainty thing as Carla Reins. Holding his head up proudly, he moved forward with a grim look on his face.

"I thought you were going to pass out back there," she prodded him. He was not amused and glowered at her, eyes growing smaller as the coppery brown skin around them constricted.

"Real funny," he grumbled. Wheezing, heavily affected by the harsh sun, he grabbed his own monstrous canteen and drank down several swigs. When he opened his mouth again, he exhaled several times, venting off some heat from inside his body. "I need a break; we can take some shade up by those rocks." Carla followed the direction of his pointing finger, turning toward a prominent rock formation.

She followed him up until they reached the solitude of the shadows. Escaping the harsh sun felt wonderful. Collapsing down to the ground, the duo took a much needed rest. Though the air temperature was still blistering, the shade cooled the environment of their harsh journey by several degrees. As the companions took refuge from the climate, a denizen of the desert caught sight of them.

A lizard with a blue ridge of sharp scales upon its back rushed from a nearby rock. It leapt and landed near Globulus, eyeing him with a pair of unblinking eyes. Looking back, the beastly hippo regarded the creature with a bemused stare. Cocking its head from side to side, the lizard considered Globulus. When he grew weary of the duo, the reptile flicked its tail several times before rushing off into a crevice within the rock face. Smiling at the lizard, Carla curled her knees up against her chest, resting her chin upon them, and then turned to Globulus with a sullen look.

Returning her gaze, he could tell she was exhausted. Usually, she rambled on incessantly about everything under the sun. Currently, she was distant. Globulus had grown accustomed to her wild shrill voice and gift for gab. Even though he would never admit it to anyone, not even himself, the hippo enjoyed conversing with her. It was the first time in his life that he had a real friend, someone he could confide in. Seeing her exhausted and quiet made him feel that things were out of place. He took a brief moment to consider his options. Starting a conversation with her could be downright annoying at times, but currently, he was missing the companionship. A war waged in his mind whether to unleash the floodgates of Carla Reins' endless chatter. Finally sick of the silence, he took the plunge and spoke to her.

"So what are you going to do after all of this?"

She was silent and looked at him, seemingly stunned as his words traversed the neurons in her brain. It was uncharacteristic of the harsh, brutish hippo to engage in conversation. Feeling like he was leading her into some sort of trap, she eyed him carefully before responding in a mystified tone, "What am I going to do after all of what?"

"You know, all of this, the war," he responded.

Blinking several more times, she shook her head in confusion, unable to comprehend that he had actually initiated a conversation. "Seriously?"

Feeling awkward, he grumbled, "Just forget it."

"No, wait," she recanted. "I'm just surprised, that's all."

"Surprised about what?" He eyed her with disdain.

"This is the first and only time you have ever started a conversation. Even more amazing, you asked *me* a question."

"I promise never to do it again."

"No... don't be like that. It's a compliment. You are becoming a civilized hippo." Giggling, she looked at him. The fire had returned to her eyes. No longer was she marked with exhaustion. Carla had sprung back to life.

"I am tired of all of this: the chaos, the bloodshed... I want it all to end. If I can live to see that day, I can count myself lucky," Globulus told his companion, now beaming a broad smile toward him.

"I never thought about what I really wanted to do. I have been out here, in the wilds, for so long; I never even imagined a day when life would be different."

"You must want to do something. Everyone has some sort of goal or ideal."

"Everyone has some sort of goal? What about you, then? What do you yearn to do after all this bloodshed is done?" Carla quizzed the mighty hippo warrior.

"I want to go home," he responded immediately. "I want to live with my people, in the marsh."

"Having a home would be wonderful." Her gaze grew

distant. Young Carla had no home. Her entire family and town no longer existed. She was a solitary traveler, a drifter, a wild spirit.

"What about you? The last thing I want to see is you creeping around the wasteland with your sniper rifle killing things. You should find a home and start a life, a real life."

"Real life? What is that anyway? Putting down roots?" Her tone was sarcastic.

"You know what I mean. This is what you want? A never-ending killing spree? Being a hired gun? That isn't you."

"You're that certain?"

"Yes, I am. You are no mercenary, only a victim of circumstance," Globulus responded.

Turning away, she grew silent. In her heart, all Carla ever wanted was a home and a family, some place to settle down and live a simple life. Globulus was right; she was not a hired gun at heart. The life of a mercenary, hunting big game for money, had been convenient. Being drawn into the war was also convenient, a way to avoid what she really wanted: a peaceful home. But Carla was tired of convenience. She hungered and longed for something more. Globulus had brought her desire to the surface. "I can't do this forever. Some day, I will break free of all of this. I will settle down in some wasteland town, live out my life in peace. You know me too well." She sighed. "Too well…"

Breaking the tension, Globulus changed the subject. "We should get moving. There are only a few more hours of sunlight left, and I want to get closer to that watering hole you keep telling me about."

Looking up at him, she smiled. As she stood up once more, Carla was refreshed in both body and soul for the time being. "You're right. I think if we push hard for the next few hours, we can make it by dusk."

And so the companions set out once more. Their goal was to reach the Iron Kai controlled territory, a long and arduous journey. Globulus had been anxious to rejoin his tribe, and his tribe would undoubtedly be equally anxious for Globulus to return. After the death of their leader, Chief Stoneskin, it was Globulus' right and honor to lead the tribe. He needed to get back to them soon and guide them through the strife of war.

Still many moons from the northlands, the duo would have to

traverse the heart of the wasteland, hopping from town to town on their journey north. Currently, they were in the middle of nowhere, over thirty miles away from the next outpost. The enemy-occupied territory through which they traveled was also rife with unseen menace. The Reaper Kai were known for sending scouting parties across their occupied territories, ready to capture any spies or hapless travelers who got entangled in their sinister web. Carla and Globulus were flies, trying to quietly pass through the web spun by the dread Reaper Kai, spanning all the way from the Steel Crag Mining Guild to Rasheed to the southeast.

The two were making excellent time, refreshed by their late day siesta. They had only pressed a few miles from their resting spot when they came upon something unsettling, something that turned their blood to ice even in the blistering heat of the day.

Globulus and Carla charged in upon the frightening scene quickly; the mighty hippo wanting to view the sight first hand. A swath of earth had been thoroughly torn apart. The cracked clay had been crushed and disturbed by the passing of thousands, a legion of troops. Tracks were pressed into the brittle clay earth, the tracks of thousands of Biogtechs on the march.

Crouching down near the swath of disturbed earth, nearly a half a mile wide, Globulus traced the sinister footprints of the Biogtech soldiers with his fingers. Shaking his head in dismay, he looked up at Carla with a look of concern.

"Tell me, young Carla Reins, where in the hell did all of these tracks come from? We are in the middle of nowhere. Why would the Reaper Kai be moving thousands of Biogtech troops through the middle of the open desert?"

"I'm not sure. I have been in this region dozens of times. There is nothing here, absolutely nothing out here."

"The tracks are heading southeast, toward Rasheed. The Reaper Kai must be reinforcing the city. But it doesn't make any sense. The Iron Kai have completely cut off troop movements out of Detro Tech City. Where the hell are all of these soldiers coming from? The tracks lead in from the northwest. The garrison at Rust Spire is southwest..."

"To the northwest is an expanse of shifting sands known as Metal-Weaver Flats. There's a small town only a day's journey away, and after that, there is nothing but open desert. A small chain of mountains lies to the north; other than that, there's nothing. Not a single outpost or town is up in that region. It is essentially a dead zone. Not even a single plant can survive in that hell-blasted patch of earth."

Shaking his head, Globulus stood up and walked into the center of the swath of footprints. Staring around him, he was awestruck. "How many Biogtech soldiers do you think made this? Several thousand at least? Maybe more?"

Her sense of apprehension made the hair on Carla's neck stand on end. It didn't make any sense. How did thousands of Biogtech soldiers appear in the middle of the open desert, seemingly out of nowhere? "I'm not sure," she answered. "I would say at least two thousand, maybe more."

The hippo warrior fell silent. His eyes moved toward the northeast, the direction of safety, the Iron Kai territory, then shifted to the northwest, the direction of danger, where the trail of tracks originated. Indecision was upon his face. Seeing his distress, Carla came over to him and they stood in the middle of the swath of enemy tracks.

Considering their options, he knew they had to make a tough decision. "We have two choices, Carla. We can head northeast and to the safety of the Iron Kai empire, or we can push northwest and follow these tracks back to their source."

Carla's brow furrowed in a look of dismay. "Some choice," she whispered. "You know we don't have a choice. If we don't investigate where these came from…" her hand made a sweeping movement over the swath of enemy Biogtech tracks, "…the Iron Kai will not be able to survive the war with the Reaper Kai. You said it yourself on many occasions. The Iron Kai can fight a war against the Reaper Kai only if they are contained. Obviously, they are not contained. We don't have a choice, Globulus. We have to follow these tracks back to their source."

"I know…" he whispered back. "If the Reaper Kai have a secret production site, *we* need to find it."

Grasping her shoulder, he looked at her grimly. She returned his gaze, knowing exactly what he was thinking. Following the enemy tracks back to their source was going to be dangerous, extremely dangerous. Instead of heading to the safety of the Iron Kai empire, the last bastion of freedom, the duo would head into the lion's mouth and hope with all their hearts that the lion wasn't hungry.

With a hesitant step, Globulus moved toward the northwest, following the trail of enemy troops emanating from the open desert. Carla followed, grabbing her rifle and checking it as they walked. Pulling the bolt back, she inspected the chamber nervously. A round slid back, exposing the primer. Her hand touched the bullet and she let out a large sigh. Back to the fray once more; back into danger again. Her anxiety building, she slammed the bolt back into place on the weapon. She gazed to the northwest and beheld the tracks left by an army of enemy troops with a tense, frightened expression.

"You ready for this?" Globulus quizzed his companion.

Staring back silently, she gave him a half hearted nod.

Step by agonizing step they moved, further from safety, closer to danger. Walking the lonely path through the wasteland, they followed the tracks laid by enemy soldiers, cresting toward the most horrible secret that the Reaper Kai empire had yet to unveil.

Chapter 32
Behind Enemy Lines

The buildings smelled of death. Since the enemy invasion several months ago, the hapless victims of the war had already spoiled to a horrid state, rotting in the hot air and harsh climate. The dead town was becoming an all too common sight in the region. With the Reaper Kai grinding their way across the entire civilized region of the Darken Realm, many towns had been utterly laid to waste, just like this one.

Bright light radiated from the sun, bathing the desert town in hell-blasted heat; simultaneously, a cool draft from the highlands rolled into the area, which lay near the shadow of a great mountain range to the north. This descending blast of air rushed off the mountains, driving back the burning heat of the wasteland. The town's location caused the chilled mountain winds to blow directly down the main street. These icy winds blasted the town constantly. Due to this unique climate, supplying a reprieve from the harsh heat of open desert, many inhabitants immigrated to the town, making it one of the largest in the area.

A barn constructed of worn, gray timbers moved under the northern winds. The enormous barn doors swayed in the breeze, their tarnished hinges whining as the doors were blown back and forth. Other buildings moved and danced in the wind, seemingly of their own accord. Shutters banged against their wooden frames. Curtains, stained with past conflict and grime, billowed outwards, dancing like ghosts in a cemetery. A screen door wiggled back and forth to the rhythm of the drafts. Dust devils rolled down the middle of town, swirling quicker and quicker, dragging plumes of reddish dirt and dust into tight vortexes. For a ghost town, there was movement all around, haunting reminders of the long-gone inhabitants.

In this tattered place, amongst the dead, was where a loathsome creature lay in wait. Making its home in the large barn near the edge of town, the beast lingered in the solitude of the empty buildings. Driven into madness by its lonely existence, the being's shattered sanity had turned it into a creature of instinct alone.

And so, sheltered in the barn, it feasted on the carrion of slain animals and human corpses. Perching on all fours, it sniffed at a half eaten horse. Its snout was black and its maw stained in cold, rancid blood. Black beady eyes gazed hungrily upon the carrion feast. As the smell of death wafted into its nostrils, it pulled its tattered clothes about it. Where once the beast had worn splendid crimson robes, now only torn remnants of cloth clung to its body. Soft ears covered in black fur protruded from its head. Gnashing bloodied teeth together, the creature threw its head back. Baying at the rafters in the barn, the crazed Goat Minion bleated. Its chin quivered frantically as tufts of its fur stood on end, wild eyes scanning the top of the barn. The rafters in the barn did not respond; the loathsome beast was alone, still alone, as it had been for months and months.

Suddenly, its ears twitched. Something at the edge of its hearing caused the creature to tense. Holding its breath, it waited as its ears moved back and forth. A thin, weak sound once again perked the senses of the Goat Minion. Madness flashed in its eyes. Leaving the side of the fallen horse, it moved across the ground on all fours. Its hind legs clicked against the floor, hooves smacking the ground with sullen noise. Bleating in joy, the Goat Minion knew that it had company. Rushing out of the barn, it slunk against the ground, searching the ghost town for the intruders it had sensed with its sensitive ears...

Panning the territory from edge to edge, the sniper scope passed back and forth, sweeping the ghost town for any signs of life. The buildings were silent, as the town had been abandoned for quite a while. Breathing softly, Carla could not see anything in the town ahead. Bringing the mighty gun down from her shoulder, she gestured to Globulus, who saw the hand signal and moved toward her position.

On a small ridge, just south of town, the duo prepared to advance upon the shell of civilization. Moving closer to dusk, the sun touched the edge of the mountain range. Slinking down, the edge of the glowing ball of fire licked the mountain top and began to set. Orange flames engulfed the clouds. Purple haze and tendrils of soft light sprang forth as the sun issued its goodbye. Long shadows erupted, bathing the ghost town in an inky darkness. The temperature dropped almost immediately after the sun set. Growing colder by the minute, the desert ghost town was conquered by the chill winds which blasted down from the mountains, howling through the middle of the abandoned settlement, their fury unmatched.

Clouds of dust rushed down the main street, creating a wispy haze of red grit. An ominous rush of sound echoed as the wind battered the silent buildings. The doors slammed open and shut; the shingles upon the houses creaked as the wind tried to tear them free. As they listened to the flurry of sound, an eerie feeling washed over the two companions.

"There is so much noise..." Carla whispered, looking at Globulus. "But even with all the noise, it feels so lonely here. There is no laughing of children, no barking of dogs. This entire town feels like a crypt, a noisy crypt."

"The tracks are almost gone." Globulus was referring to the thousands upon thousands of Biogtech tracks left in the desert. The duo had been trying to follow the tracks back to their source. "The wind is destroying any sign of the army's passing. We must hurry if we want to discover the source of the Biogtech soldiers." Nature was a harsh enemy, which had removed the majority of the trail already; the harsh winds hid the tracks under blowing dirt and shifting sand. Globulus and Carla had spent the past few days following a scant trail deep into the northwest desert.

"We must be careful." She shook her head in apprehension. "My blood is chilled, it feels like ice. I have a bad feeling." Rubbing her shoulders in the chill air, she shivered a few times, looking toward the ghost town ahead.

"Let's push on, it's getting dark. We can spend the night in one of the buildings ahead. It seems safe enough."

Carla looked at the empty buildings as fear filled her mind. Corpses of dead citizens lay in plain sight, strewn about the avenues. What did the silent homes hold? What sinister horrors were hiding inside the buildings? Driving the hideous thought from her mind, she prepared to approach the town with her companion.

The town was arranged in a grid pattern, sprawling over a dozen blocks in every direction. Globulus led the way as they entered the southern edge of the city, his eyes constantly swiveling back and forth. The shadows grew more intense as the sun passed further beyond the horizon. Squinting, he halted as a harsh blast of chilling wind struck him. A plume of grit stung his eyes. Shielding his face, the mighty hippo took a brief moment to reorient himself after the plume of swirling dust passed by.

Rubbing the dirt from her own eyes, Carla caught sight of something at the periphery of her vision. As she spun around, something reddish in tint disappeared behind an abandoned house. Bringing her rifle to her shoulder, she breathed erratically, frightened by what she had seen.

"What is it?" Globulus quizzed, drawing his brace of shotguns.

Moving forward with her weapon at the ready, she approached the edge of the building. Walking a wide arc around the periphery, she moved methodically, her steps confident. Globulus was beside her, holding his own weapons ready.

"Let's do it," he grunted. Moving quickly, they both rounded the corner with weapons set for battle. But nothing awaited them beyond the corner.

Breathing a sigh of relief, Carla smiled and let out a dry laugh. "Must have been my imagination."

"Yeah, you're probably just rattled. Our nerves are running a little rough and I can't blame either of us. Tracking an entire army is risky business, especially through the ruins of battered towns. With the sand stinging our eyes, it's easy to see things that aren't there, especially in this dim light."

"Yeah, my eyes are playing tricks on me. It's been a long day. Let's find a place to stay and bed down for the night."

Darkness had fully engulfed the wasteland. The soft hues of

the sunset had been replaced by icy blackness. Locating an empty saloon, the duo moved inside, looking for a place to make camp in the eerie ghost town. Using their flashlights, they peered into the interior and found the place in shambles. Tables had been smashed and overturned. The bar had been ransacked. As they passed into the saloon, broken glass crunched under their combat boots; shattered liquor bottles covered the floor. Taking a quick survey of their surroundings, neither of them saw anything dangerous, and they decided to make camp. They unpacked their gear, rolling out their sleeping bags near the back wall.

Globulus turned off his flashlight and began to rummage through his pack, finally finding a small kerosene lamp. Carla aided him by training her flashlight on the device. He lit the wick, which began to burn, emitting a dull yellow glow. Turning off her own light, she sat on the floor.

The wind howled outside the saloon. Whistling through a broken window, the rough air whipped into the room with a haunting wail. Pulling her knees close to her body, Carla listened to the wild wind. The building seemed to groan, lurching and popping rhythmically. It almost sounded as if something was on the roof. Creaking, the worn boards flexed in the strong wind. The entire building seemed alive; the wind rushing through with a whisper, boards creaking, curtains waving back and forth. Trying to repress their eerie surroundings, she closed her eyes and thought about something else.

A vision of the mountains filled her mind. The imagery was pleasing. For a brief moment, she drove away the sullen surroundings and thought about splendid snow fields between the mighty mountains. She imagined being up in the snow, watching the spring cascade; pure, pristine water was crashing downward from a great height.

The anxiety of the strange ghost town was beginning to abate. Rolling onto her back, she rested on the sleeping bag. Laying her rifle against her chest, Carla's left hand began to rummage through her backpack. It emerged with a lump of jerky and some stale bread, and she turned to eat her nightly meal.

Globulus was silent, also listening to the eerie sounds of the harsh wind battering the ghost town. He rose to his feet and stared out the window of the saloon. Clouds of dirt were obscuring his vision. The dust storm was so thick, the hippo warrior could not even see across the street.

"I'll take first watch," he said in a commanding tone. Nodding even though he could not see her, Carla silently agreed. As a wave of sleepiness washed over her, the young woman blinked several times, her vision blurring. Turning off the kerosene lamp, she drifted into a deep slumber...

Slam! The front door to the saloon was hurled against its wooden frame. Jolted back from her slumber, Carla sat straight up in her sleeping bag. Grasping her weapon in shaky hands, she looked around in the darkness.

"Globulus?" she whispered, trying to locate her companion.

No one responded. The wind rustled through the curtains with a hiss.

"Globulus?" she repeated in a quicker, more frantic tone.

Still, no one responded. Somewhere in the saloon, the shattered glass on the floor grated against the wood. Something was moving, near the window. Thinking that her ears were playing a trick on her, she held her breath. A split second later, another piece of glass cracked under strain, popping with an ominous sound.

When her left hand groped toward Globulus' sleeping bag, she found it to be empty. The mighty hippo was nowhere to be found. Panic took over Carla's senses. In the darkness, a floor board creaked under an unknown weight. The panic intensified to terror. Fumbling in the dark for her flashlight, she grasped it quickly and turned on the beam, panning it from one side to the other.

Near the window, the beam exposed a hunched form, garbed in shredded red robes partially covering swatches of black fur. The creature's head was hidden by its tattered hood. Ice filled Carla's veins. Her arm spasmed uncontrollably. The flashlight fell from her limp wrist and hit the floor with a clatter, extinguishing the beam.

"Help me!" she whispered.

The floor creaked again. Something was breathing in the darkness.

Bringing the bolt on her rifle back, she ensured that a round was chambered. The shell clicked into place. Straining her ears, she retreated across the floor until her back was against the wall.

The cloying darkness hid the demonic menace. Sniffing the air, it probed the darkness for her form. Her fear and her scent filled its nostrils. The more it sniffed, the more its madness drove it toward violent feelings. It wanted to hurt her. It wanted to catch her and smell her whimpering in the dark.

Creak... It moved forward, hunting her in the darkness. Trying not to scream in terror and give away her position, she panned her weapon toward the sound of the creaking instead. Backing along the wall, she tried desperately to find an exit.

Anger flooded the beast. Evil was coursing through its veins. Driven by madness and a need to kill, the Goat Minion lumbered forward into the darkness, searching for young Carla.

More broken glass crackled on the floor of the saloon. The fear spiked within Carla. She knew that at such close proximity to the creature, she would only get off one shot. After she fired her weapon, the beast could find her instantly in the darkness. Knowing that patience and wild luck were her only chance for survival, she continued to skirt the edge of the room. If she was lucky, Carla could make it over to one of the open windows and escape into the street, where she had a better chance of survival.

Scuttling backwards, her left hand pressed down on the floor and was pierced by a shard of broken glass. Pain bolted through her as the shard cut her flesh. Exerting all her self control, she managed to refrain from shrieking in pain and fear, even as blood dripped from the open wound.

Blood, sweet blood, filled its nostrils. Sniffing the air like a lion, it moved towards the smell with frightening speed. The scent was wonderful: fresh blood and fear. The Goat Minion began to breathe heavily, driven into a frenzy by the smell of her blood.

The wooden floor creaked once more as it rushed towards her. Hearing the sound of motion, she trained her gun in its direction. As she held her breath, she could hear the creature breathing, even in the harsh wind. It was close, very close, maybe even only a few feet away, a mere arm's length away.

Her trigger finger twitched. As her blood turned to ice, she knew evading the creature would be nearly impossible.

"Globulus, where are you?" her mind screamed.

Its hoofed foot stepped forward. The scent was strong now. It could almost feel her next to it. Crazed, the beast hungered for her.

The fear and terror were even more intense as she felt her heart pounding in her chest.

A blast of dust rushed into the saloon. A huge form, monstrous in size, had entered. The Goat Minion whirled around in the darkness. Sniffing the air, it took a step back. The situation was no longer in its favor. The hunter would only seek its prey if it had a sizable advantage. That advantage was gone.

Globulus felt alarmed. His own powerful sense of smell had told him *something* other than Carla was inside the saloon. Grabbing a shotgun in one hand and his flashlight in the other, he flicked on the light, which fell upon the sinister Goat Minion. As the beam of light hit the minion of darkness, Carla immediately trained her rifle on the creature. Taking hasty aim, she fired.

The heavy bullet tore forward and slammed into the beast's shoulder. Piercing flesh, the round slammed through, and a spray of blood erupted from the wound. Screaming in agony, the beast stumbled back. Its black eyes wreathed in pain, it followed the beam of light. The beast's madness drove it into a killing frenzy. Pulling a jagged, gore-stained knife from its belt, it emitted a high-pitched war cry. Bleating and baying, it charged the hippo.

Globulus was stunned by the attack. Palming a shot gun like a pistol in his enormous hand, he pulled the trigger. The scatter shot blasted forward and struck the crazed Goat Minion. Undaunted by the further damage to its body, the frenzy in its heart grew. It wanted to kill, more than anything else.

Globulus fired his semi-automatic shotgun repeatedly into the charging creature. The mighty hippo warrior managed to hit it three times during its wild advance. Unfazed by the attacks, the Goat Minion jammed its knife into him. The blade tore into the hippo's right thigh, plunging deeply. As he reeled in pain, the Goat Minion never wavered; it pulled the blood-stained knife free and continued to stab him relentlessly. Globulus was cut several more times, and the damage to his leg was becoming severe. Even his thick hide was no match for the vile will of the creature.

Raising his weapon to defend himself, Globulus leveled the weapon at the beast's head. He fired at point blank range; finally, the blast was more than enough to stop the threat. The creature took the brunt of the shot in its face. Falling to the ground, it died immediately.

Seeing the creature fall, Carla ran forward to aid the injured hippo. He had dropped to one knee, holding his bloody leg and wincing in agony. Cursing, he pressed both of his hands near the wound, applying pressure to stop the bleeding. It took about a minute for the wounds to stop bleeding profusely. Wrapping a shirt around the wounds, Carla fashioned a makeshift bandage.

"Are you all right?" he grumbled, more concerned about his companion than about his own condition.

"Yeah, don't worry about me. Are you all right?" she asked in apprehension, rubbing some of the blood from his leg.

"For now." Crouching down, he grabbed the Goat Minion's gore-spattered blade. The weapon was grotesque, covered in the remains of dozens of other hapless victims. It didn't take a medical doctor to figure out that the weapon was coated in rotting flesh. His wounds were definitely going to get infected. "Find something to wash my wounds out. Hopefully there is something still left in this bar to clean them out."

Going behind the bar, she found a bottle of liquor which she used to wash out the wounds. As she witnessed the severity of his primary injury, she shook her head in despair; it was deep, really deep, and with the knife covered in rotting flesh, he was sure to get an infection.

With the threat posed by the Goat Minion now ended, their real concern would be the hippo's health. As they were stuck in the middle of the wasteland, weeks away from any medical aid, his life would teeter on his ability to withstand infection.

Helping him over to his sleeping bag, Carla tried to support part of his weight. They sat down with a thud, hitting the floor of the saloon heavily, then turned to each other with serious gazes.

"I suggest we both keep watch the rest of the night."

Carla agreed with a nod. Turning off the flashlight, the two companions let the sounds of the night flood them once more. The wind howled in the darkness, and timbers creaked and groaned. Eerie sounds invaded their minds. Staying the rest of the night in the saloon would not be easy.

Chapter 33
Metalweaver Flats

With a gasp of pain, Globulus took another labored step forward. He was tough, but even the seasoned warrior's enthusiasm had faded. With pain filling him and a faint dizziness beginning to take over, Globulus knew that he was running a low grade fever and that the wounds he had sustained when battling the Goat Minion were becoming serious. "I need to take a rest."

Carla looked at him with concern. The duo had been traveling several days since the incident in the ghost town. Their pace had dwindled to nearly half their usual speed, as Globulus was increasingly unable to move quickly. "Agreed. Let's rest over there, along the rocks," she responded.

Coming to rest in the shade of a rock outcropping was a welcome relief from the heat. It was afternoon, and the companions had taken the sheltered passage through the valley, preferring to keep to the foothills to avoid being spotted by enemy scouts. Earlier that same day, they had eluded a Reaper Kai scouting party that was making a pass through the valley. They had also come across the tracks of two more Reaper Kai tracker teams just after noon. This tidbit of information was intriguing. Whatever was in the region, it necessitated a heavy enemy presence, which further piqued the companions' interest. Whatever was hidden out in Metalweaver Flats, Carla and Globulus had resolved to discover it.

"Let me have another look at your leg," Carla ordered, and the mighty hippo nodded in agreement.

As she pulled back the bandage, the smell of filth wafted

forth. Reeling back from the wound, Carla held her breath for a brief moment. Globulus was not pleased by her reaction.

"I think it's getting worse," she said in dismay, not pulling any punches. Moving closer and breathing through her mouth, she inspected the injuries. The two smaller knife wounds had already closed and were looking well on their way to healing. The larger of the puncture wounds, however, was swollen at its edges. Placing her hand on the swelling brought forth an eruption of pus. Wanting to purge the remainder of infection, she pressed hard again. Another thick mass of white pus erupted. Globulus gripped his hands together as a jolt of pain rolled through his leg.

"Don't push so damn hard!" he grumbled.

Ignoring his pleas, she pressed again, forcing the remaining pus from his leg. After it was purged, she wiped away the remaining traces of infection. "Get ready, this is going to be painful." Young Carla was referring to the bottle of liquor she had just removed from her pack. Using her fingers to pry the wound open, she exposed the infected flesh. Taking a part of the bandage, she placed it into the wound and began to wipe away the yellow pus.

"Argh!" Globulus yelled in pain, trying to maintain his composure as she worked on him.

"Just a little sting now." Still holding the wound open, she poured some of the liquor directly into it.

The pain was intense. It felt like fire burning every nerve ending in his leg. Nevertheless, he remained stoic, and didn't even whimper. It was amazing to see him withstand so much pain. His grimace stayed in place for several seconds, as the pain subsided. Opening his eyes, he pulled in several quick breaths of air. With the burst of fire from the alcohol dissipating, the only thing that lingered was a dull throbbing pain.

"We need to stop and clean it out more often," Carla said firmly, her expression concerned.

"We don't have time for that," the mighty hippo grumbled.

"Time? If it gets any worse, I'm gonna have to saw your damn leg off. Don't worry about time. Worry about your leg."

"We can't just wander around out here. You saw that enemy patrol before, and we have seen signs of two other scouting parties passing through here recently. The enemy, for whatever reason, has a strong presence in this region. If we move any slower, we will be caught, and our own little scouting expedition will end quickly."

"Maybe I should scout ahead, and come back for you? You can hide out here in the rocks. It will be better than walking on that infected leg of yours."

Sighing, he shook his head in disagreement. "No, we do this together. I am more worried about your safety, pushing on alone, than my own safety going with you."

"Come on, I could be back in a few hours. With as many enemy patrols as we have seen, we have got to be close to *something*."

"No," he responded firmly. "You are not heading out alone."

Still undaunted, she pressed on again. "It will be just a few hours. I will be fine. I can elude them easily."

"Are we speaking the same language?" he rumbled, his face flushed with anger, opening his enormous maw to expose tusk-like teeth. "I said no. I am not letting you go on alone. That is the final word. I am going with you, and we are leaving right now!"

Her eyes widening, she fell silent. Globulus had finally done a good job at shutting her up. Nodding in agreement, she grabbed her gear and stood up. Pulling on his arm, she helped the enormous hippo hybrid to his feet.

With silence between them, they set out once more to discover the secret of the enemy soldiers hailing from the heavily patrolled, enemy-occupied valley. The trail of troops had faded to a mere hint of their passing. As winds and shifting sand were the primary terrain of the region, even the smallest breeze could wash away any sign of the army's passing. The duo was fighting a failing war against the very environment. If they could not locate the origin of the army soon, nature itself would eradicate any helpful clues.

In order to avoid enemy scouting teams, Globulus and Carla had skirted the sands, preferring the rocky region closer to the base of the mountains. The imperative not to leave their own set of tracks was another primary concern for the companions. So far, they had done an excellent job of evading the enemy.

"I haven't been out here in about three years," Carla said, wanting to break the tense silence between them.

"I see little reason to ever come out here. Other than that ghost town a few days back, there is nothing out here, and I mean *nothing.* Why on earth did you ever come out here?"

"I thought it would be a good place to hunt. Sad thing is, not even the big game of the desert frequents a place like this."

"I can see why."

In the shadow of the mountain, they moved in a northwesterly direction. Above them to the west was a set of steep foothills. Being so low in altitude, the foothills did not receive snow like other, more prominent mountains in the Darken Realm. This environmental factor made the range extremely arid. With no run-off of snow, the valley was a scorched, lifeless swath of earth. Only rarely did the duo come across anything alive. From time to time, they noted a cactus or a yucca plant growing out of a shadowy crack in the rock; these were the only forms of life that could dredge out an existence in the inhospitable climate.

Still walking with labored movements, Globulus plodded on in silence. Carla could tell he was in pain, but the rugged hippo was too tough to ever admit it. She worried about him immensely. They had been traveling together for a long time now, and they were good friends. It was hard for her to watch Globulus slide toward ruin as the infection continued to take hold of him.

They were both lost in thought and had fallen silent. With each move Globulus made, his thoughts focused on the pain and his will to continue. Carla was dreamy eyed, her thoughts drifting from her concern for Globulus to fantasizing about the day the war would end. Her ruminations were diffuse and somewhat scattered. The continuous travel through the heat was exhausting both of the companions, and her mind was beginning to wither as her fortitude for travel waned. And so they moved, quietly, deeper into enemy territory and deeper into danger.

It was about an hour from dusk as they reached the end of the valley. The terrain was getting rockier, ascending up the side of the mountain. Passing over a small ridge, they found themselves exposed in relatively open territory, and were quickly set on edge. On the other side of the rocky pass, just beyond the boulder field, was a sight that would forever chill their souls.

Guided by anxiety and a sense of self preservation, Carla dropped down to one knee, as did the hippo. Quickly, they concealed themselves behind a chain of boulders and stared at the valley in disbelief.

Below them was an enormous excavation site. Thousands of Biogtech soldiers and hundreds of Reaper Kai were amassed in the valley. Heavy loaders and other machines were pushing sand away from the center of the gorge. The shifting sands had smothered a massive industrial complex built by the ancients, hundreds of years ago. The Reaper Kai were in the process of unearthing it.

Already, over a dozen buildings had been cleared away. The complex was buzzing with activity. Conveyor belts were churning at full speed as rock and ore from the mountain were being funneled toward the facility. The rough ore was moving into a processing facility via the conveyor belt. Using the raw materials from the mines, the Reaper Kai had the capability of producing steel.

The north end of the complex contained an enormous series of towers. Pipes jutted from the side of the structure, running several hundreds of feet into the air. Steam and black smoke were billowing from several stacks on top of the building. The building was an old world petroleum refining plant. Currently, the refinery was churning out distilled fuels to be used either for the site's energy needs, or to synthesize plastic polymers and other chemicals.

Old world laboratories and workshops had also been uncovered, giving the enemy additional applications of ancient technology. A priceless wealth of knowledge was being deciphered by the sinister Reaper Kai scholars and technicians. The complex was only about halfway unearthed, and the unmined potential for old world technology was staggering. The facility was a treasure, one of the biggest finds that the Reaper Kai had discovered thus far. A functioning nuclear weapon was the greatest prize an army could find in the ruins of America, but the production facility in Metalweaver Flats was a close second.

Shaking their heads in dismay, the companions viewed the massive industrial complex, using binoculars and Carla's gun scope. As a sense of dread crept into their hearts, they knew that this discovery would be vital to the survival of the resistance against the Reaper Kai.

"Look over toward the north end of the complex. See all those soldiers?" Carla whispered in awe.

An enormous courtyard was set in the center of the facility. Several hundred pale Biogtech soldiers were standing, rocking back and forth, their mindless computer processors waiting for orders. A

Reaper Kai priest wheeling an enormous cart filled with guns was going from Biogtech to Biogtech, outfitting them with deadly weaponry. They accepted the dangerous weapons with blank expressions.

Their surveillance of the compound was increasingly chilling. The Reaper Kai had unearthed an enormous foundry and they were producing hundreds, if not thousands, of Biogtech troops, secluded in the mountains. If not kept in check, the production facility could easily produce more troops than the Iron Kai could handle.

"I think the mystery of all of the Biogtech soldiers has finally been solved," Carla said somberly, still looking through the scope on her rifle. "The enemy isn't smuggling soldiers out of Detro Tech City. Instead, they are being produced out here, without any military pressure. It makes sense now. The Iron Kai have blockaded the Reaper Kai capital for many months. The army that sacked Rust Spire did not hail from the northlands. Instead, all of the troops were produced here, in this valley. I cannot believe it hasn't been discovered until now."

"We have got to warn the Iron Kai. Without their intervention, the war will be lost."

"We are over eight hundred miles from Iron Kai territory," Carla whined in dismay. "Even worse, neither of us has a radio transmitter powerful enough to travel more than twenty miles. The enemy has control over this entire region. With the Steel Crag Guild destroyed and Rasheed under the enemy's control, there is no ally in this region that could even help us get a communication to the Iron Kai."

"We have to get back and warn the Iron Kai, even if we have to walk there," Globulus said in a harsh tone. The gravity of the situation was formidable. If they did not warn the Iron Kai command, the Reaper Kai would continue to produce thousands of Biogtechs unchecked. Eventually, they would overwhelm all that remained in the Darken Realm. "If we cannot get this information to them, the Darken Realm is doomed."

Carla and Globulus had a new mission, a mission that, if they failed, could mean the death of all. If they could not reach Iron Kai territory and warn the unified troops, the Reaper Kai could produce thousands upon thousands more Biogtechs and eventually overrun the entire Darken Realm.

With the responsibility of this task as their primary burden, both of them knew the road home would not be easy. After seeing the massive troop build-up in the valley, the duo knew escape from Metalweaver Flats would be difficult. With Globulus' injuries slowing him substantially, the grim fate of all the realm's inhabitants was hanging by a thread. If Globulus and Carla could not break through and get a communication to the Iron Kai, the entire Darken Realm was doomed to be slowly overrun by a never-ending horde of enemy troops.

Chapter 34
A Tough Decision

"Listen to me!" Globulus said in an excited tone, grabbing his companion's shoulders and shaking her briskly. An expression of urgency creased his brow. Carla looked back, terrified, breathing erratically as the tension of their grim situation began to take hold of her psyche. "You have got to get out of here! At least one of us needs to make it back alive!"

"I am not leaving you!" she said in a whimper. Glancing over his shoulder, he saw the enemy advancing quickly on their position. Ever since they had left Metalweaver Flats, the newly discovered Reaper Kai stronghold, the duo had been trying desperately to escape the valley without getting entangled with enemy scouting parties. Despite their skill and careful actions, they had caught the attention of not one scouting party, but three Reaper Kai reconnaissance teams.

"You don't have a choice. You have got to get out of here and out of here now!" he rumbled back.

"We can..." she stuttered desperately. "We can make a stand."

"A stand? They have over twenty Biogtechs and three Reaper Kai priests. We can barely handle one scouting party, let alone three. Go!"

She wavered, gripped by turmoil. Blinking several times, Carla tried to hold back the flood of tears.

"I can't escape with this infected leg. I can barely walk, let alone run. You have to go. You have to survive so you can warn the Iron Kai. If we die here today, the free peoples will suffer. You have to survive and warn the Iron Kai of that facility in Metalweaver Flats!"

"I'm not leaving you!" she whimpered. The battle to hold back the tears was lost. As they streaked down her face, she shook her head in despair.

The scouting parties were dispersed but advancing steadily, flanking their position. Enemy troops were moving in quickly on their hidden location, and the noose was beginning to tighten. Every second counted, and Carla was losing precious time, torn by indecision. If she could not escape quickly, there would be no escape.

"Hey!" he said, shaking her once more. She turned to look at him, wiping away the tears. There was a solemn look on his face as he stared into her eyes. He sighed softly, his grim expression turning to a smile. "Listen to me. Go now and run. You can still escape. Live to fight another day. Warn the Iron Kai. Without *our* knowledge of the enemy, the unified forces are doomed."

"Please, no!" she sobbed.

"Go now," he pleaded with her.

"I can't!"

"You must. Just go. Do the responsible thing and warn the Iron Kai. Go now, go quickly."

Shaking her head in frustration, she knew that his logic and plan of action were her only real option. If one of them didn't make it back alive, the Darken Realm was doomed. With her heart screaming in despair, she clutched him tightly, saying a fond farewell. The tears upon her face were like an open aqueduct. Fighting back the emotion, she left the noble hippo's side.

Moving quickly up the hillside, under the cover of the rock formations, she tried to master her emotions. She trembled, unable to even turn around. Making haste under duress, she managed to press out of the range of the three Reaper Kai scouting teams which were on the advance.

Holding his festering leg, Globulus winced in pain. Placing all of his weapons on the rocks in front of him for easy access, the noble hippo hybrid prepared to make a desperate last stand. There was no way he was going down without a fight.

"I can sense a presence..." one of the red robed priests

declared with a rasp in his voice. His psychic abilities were stretching outward, probing the area for signs of life.

"The tracks lead up there." Another priest pointed up the hillside, toward the rock formation concealing Globulus.

"I think there is another..." The third priest was also probing their surroundings with his demonic powers, picking up on Carla's wild emotions. Her distressed emotional state was a beacon for the priests. "Advance!" the priest boomed, and the mindless Biogtech robots moved up the hillside. Shambling forward slowly, the pale plastic robots giggled.

Gritting his teeth, Globulus forced himself to rise, his festering leg stinging. It felt as if fire were exploding through his leg as he rose. Fighting back the pain, he readied his weapons, preparing to fight to the death.

A twinge of anger rose in the lead priest. Having mastered his dark powers, his senses were perfectly attuned. Knowing full well that *something* was preparing to ambush the Reaper Kai forces, the priest dropped into a meditation, focusing on the hidden threat. The priest located the essence of Globulus, his will guided by sinister powers. "There you are!" he hissed. Channeling unseen energy, he began to flood Globulus' mind with powerful suggestive thought.

A drowsy feeling came over the hippo warrior. Steadying himself by placing his hands on the rocks, Globulus felt as if he was about to lose consciousness. The demonic will of the priest was forcing him to fade toward the dream world. A voice stung in his mind.

"*Sleep...* " the voice whispered.

Shaking his head back and forth, he tried to block the suggestive thoughts from his own. Though strong in battle, his mental fortitude was not sufficient to stop the hypnotic suggestion planted directly into his brain by the sinister Reaper Kai priest.

"*Sleep...* " The suggestion stung his mind again, this time much more intense.

Unable to combat the invasive suggestion, Globulus dropped to one knee, the weapons falling from his hands. Shaking his head as he blinked several times, his pulse began to feel like sludge. Fatigue was filling his mind.

"*Sleep...* " Once more, the suggestion struck his mind.

Finally, his fortitude failed.   Crashing to the ground, Globulus of Rasheed fell into an unnatural slumber.   Completely defenseless, he had been defeated by the powerful will of the demonic priest without a single shot being fired.

"Got him," the priest rasped, opening his eyes, staring up the hillside.   "What about the other?   Can you still sense another presence?"

One of the other priests was meditating, his powers probing the rock formations up the hillside.   A spike of emotion flooded the priest's mind.   It was weak and somewhat distant, but there nonetheless.   Focusing on the emotion, a twinge of sadness entered his mind.   Whoever was hiding had been gripped by sadness and despair.   Functioning as a psychic compass, the priest began to isolate the location of the hidden person in the rocks.

Whimpering, Carla was still gripped by mental anguish.   The sniper had witnessed the defeat of Globulus with the aid of her rifle scope.   Shocked by the futility of his last stand, she had been overtaken by fear.   The emotion was strong enough for the priest down below to track her.

Her shaking turned to a sob.   Tears rolled down her cheeks as the enemy advanced upon her.   Wracked with tense spasms of emotion and distress, she was losing control.   Each pulse of sadness allowed the Reaper Kai to further pinpoint her concealed position.

Knowing that both her life and the hope of the free peoples were fading, Carla understood that she had to get herself under control, and do so quickly.   With a patient, formidable will, she forced the sadness from her mind.   Breathing softly, she suppressed a final whimper as her mind began to chant.   *"I am a rock. I am a stone upon the earth."*

The priest froze, losing his hold on her emotions.   He retained only a weak, fleeting sense of her presence.

Another tear rolled down her cheek.   The image of Globulus, helpless, unconscious before a host of sadistic enemies, washed over her again.   The pulse of emotion was picked up by the Reaper Kai. Feeling her again, they advanced slowly, following her emotion like a trail of bread crumbs toward her hidden location.

Seeing them advance again, she tried to regain control.   *"I am a rock. I am a stone upon the earth."*   Forcing any other thought from her mind was difficult.   If she could not focus, the Reaper Kai would be able to sense her. With a stubborn will, Carla forced all other thoughts from her mind, thus confusing the Reaper Kai who were attuned to tracking beings through emotion.

The psychic link was beginning to fade again. All the priest could see in his mind were the rocks around him. Shaking in disbelief, he opened his eyes and scanned the formations of stone. Nothing; he could sense nothing once again.

"What is it?" another priest hissed, seeing the indecision on the tracker's face.

"I am not sure…" he stammered. "Maybe it's nothing, what I am sensing."

Carla lost her concentration once more. Her concern turned to her fallen companion.

"There you are!" the Reaper Kai rasped, feeling her presence.

Advancing quickly, the trio of Reaper Kai moved up the hillside toward her hiding place. Watching them filled her with fright, further strengthening the psychic tracker's ability to locate her. They were getting close, no more than fifty yards away.

Fighting back the flood of emotion once more, she knew that regaining her control was her only chance for survival. Using her formidable will and stubborn personality, she began to concentrate once more. *"I am a rock. I am a stone upon the earth. I am a rock. I am a stone upon the earth."*

"Lost it again!" he hissed.

"Maybe it's nothing."

"I have an idea. If there is someone up there, let's flush them out!" Turning towards the Biogtechs, the priest ordered them to open fire on the rock formations above them.

Gunfire roared forth as dozens of bullets zinged through the air, slamming into the rocks around her. The attack startled her, making her lose her concentration. The priest immediately picked up on a jolt of panic at the edge of his senses.

"We are getting closer," he hissed, following the emotion.

"I am a rock. I am a stone upon the earth. I am a rock. I am a stone upon the earth."

"Lost it!"

"Fire again!" The priest ordered his Biogtechs to fire once more.

This time, her concentration was staunch. Finding a stubborn will to live, she forced all of the sounds of gunfire from her mind. Over and over her mind chanted. *"I am a rock. I am a stone upon the earth. I am a rock. I am a stone upon the earth."*

"Troops, advance!"

The giggling death machines responded to the order. Taking shaky steps up the hill, the horde of Biogtech soldiers ascended slowly toward her position.

Climbing into a crack in the rocks, Carla wrapped her arms around herself in the secluded hiding spot. Over twenty Biogtechs and all three priests had closed in on her position and were searching in earnest.

"I am a rock. I am a stone upon the earth."

"Nothing, I can sense nothing."

"You were mistaken," another priest replied, only a few yards away from where Carla had concealed herself.

"I am a rock. I am a stone upon the earth."

"I think you are correct: I was mistaken. I sense nothing up here. Let's get back and interrogate our prisoner!" the tracker concurred, and they abandoned their search for Carla, convinced nothing was hiding in the rocks. The concealed sniper had succeeded in eluding the priests by emptying her mind of any emotion or concrete thought.

Carla continued her internal chant as the troops pulled away, turning their attention to the fallen Globulus, slumbering unnaturally down the hillside.

"What do we have here?"

The troops surrounded Globulus and secured the mighty hippo in restraints. After seeing that he was incapacitated, the head priest ordered, "Wake him!"

One of the priests awakened him with a brutal kick to the head. Still groggy, Globulus opened his beastly brown eyes and surveyed his captors. Feeling glad to be alive but dismayed by his capture, the hippo simply stared at them without any response.

"Identify yourself!" the priest rasped.

Globulus did not respond. He would never speak to such scum. Kicking him in the face once again, they tried to get him to talk.

"Speak!"

Again, no response. As they kicked him repeatedly, the hippo felt pain but was determined he would never surrender.

"Search his gear!"

Crouching down, the priests rummaged through his belongings. They quickly discovered a set of orders from Iron Kai command, addressed to Globulus. The orders were a set of maps and demolition set points within the Steel Crag Mining guild tunnels, intended to bring about the destruction of the Gold Road.

"Globulus of Rasheed!" the head priest boomed, elated by the discovery. "Your traitorous actions will gain you death!"

Undaunted by the threat of death, the hippo forced a lump of mucous to the front of his mouth. Spitting it outward, the slimy mucous struck the head priest in the face. "Do your worst," Globulus rumbled.

Enraged by the display of defiance, the head priest wiped the spit from his face. A quick death was no longer an option. A much gorier fate awaited the defiant hippo.

"You impudent, worthless wretch. You will pay for your arrogance!"

"What should we do with him?" one of the priests quizzed their leader.

"We will take him back and interrogate him, purify him with punishment and torture. After we break his will and body, we will then send his mangled body to the Iron Kai!"

Nodding in agreement, the other two priests confirmed the plan of action. A moment of silence passed between them. The head priest was viewing Globulus with fascination. Guided by a sadistic instinct, the head priest had suddenly concocted another, more sinister plan. "We could take him back to Metalweaver, or..." He smiled, almost salivating over his new twisted plan. "Take him to Rasheed and allow Queen Marion the pleasure of his torture. Imagine the agony. Imagine the suffering. I could not think of a worse fate, being tortured by your former master in your former home."

The plan was twisted, chilling, and not without irony; a perfect Reaper Kai ploy.

"Take him!" the priest ordered, and the host of Biogtechs lifted the bound hippo from the earth. With their cargo in tow, the scouting team moved away, leaving the silent desert and Carla lost in her own thoughts within it.

Breathing softly, she crawled from her hiding spot, wracked with uncertainty. Carla had to make a hard choice: either abandon Globulus to torture and certain death, or warn the Iron Kai about Metalweaver Flats. Both choices came with a heavy price. If she attempted to rescue Globulus, there was a good chance that she would also be captured and put to death; the Iron Kai and their allies would never know about the dangerous military threat, and no one would ever find out about the production facility in Metalweaver Flats. Without knowledge of the powerful production facility, the unified forces would probably be destroyed, which would ultimately bring about the loss of hundreds of thousands of lives. If she abandoned Globulus, she could warn the Iron Kai, but her friend would face a gruesome death, tortured until his body and spirit were broken.

The responsible, logical choice was to let Globulus die and warn the Iron Kai, but young Carla had the gift of compassion. No matter what the cost to anyone else, she could not find the strength in herself to abandon her companion. Choosing the reckless but compassionate option, Carla resolved to try and save Globulus from his horrid fate. With a staunch look upon her face, the sniper shouldered her weapon and prepared to follow the Reaper Kai. As she turned southward, Carla resolved to save Globulus from a fate of sadistic torture. Knowing the road was going to be long, the stubborn sniper hid her emotions deep inside herself. The road Carla was walking required fierce conviction; succumbing to emotion and weakness would only result in utter failure.

"Hang on Globulus, I will not let you down," Carla swore as she followed the sinister tracker teams holding her best friend captive.

Chapter 35
The Battle of Dakota Beach

Emperor Gunther was in shock. The radios in the Truce Hall were buzzing with communications. All of Gunther's military advisors and top officials were crowded around the communication center, flushed with worry.

Fishing vessels on the lake had radioed in that hostile landing craft, boats filled with hundreds of Biogtech soldiers, were headed toward Dakota Beach, a heavily fortified defense just outside the mighty Iron Kai capital city of Stonen.

"How many boats do you count?" Field Marshall Patrick, the officer in charge of the defense of Stonen, spoke into a radio transmitter.

"There are ten boats inbound," a loyal fisherman responded back into the radio. "By the looks of things, there are about fifty soldiers on each landing craft."

"Got it. Keep your eyes open for any more ships out on the lake." Patrick whirled around and charged across the Truce Hall to stand beside his Emperor.

"How many soldiers are inbound, Marshall?" Gunther asked solemnly. He was standing near the enormous windows that looked out on the lake below.

"We have a rough estimate of five hundred inbound enemy soldiers. They are loaded onto landing craft, and their destination appears to be Dakota Beach."

"Assemble all military personnel inside the city. Alert everyone outside the city, get them in here quick, and lock this place down tight. If this is it, the beginning of a full scale invasion, I want all of the civilians protected inside the tunnels."

"Yes, sir," Patrick replied respectfully.

"I want two thousand infantry entrenched on the hills behind the machine gun bunkers. Alert the machine gun nests overlooking the beach and give orders to fire at will. Get the artillery up and pound the hell out of the beach. Blanket the entire area with mortars and anything else you can throw at them. If there are more transports on their way in across the lake, I want that beach cleared before any Biogtech reinforcements arrive."

Patrick snapped his fingers and his aids began to give the orders just handed down by Gunther. "What about the choppers? Should we get them in the air?"

"No, leave the helicopters underground. Last thing we need is a bunch of those Biogtechs firing anti-aircraft rockets and downing our choppers needlessly. Our soldiers on the beach should be more than enough to take down five hundred Biogtechs." Rubbing his orange beard, Gunther eyed the battleground less than three miles away. The battle was going to be close, very close to the capital city of the Iron Kai empire.

"Yes, sir. Get to it, everyone! You have your orders!" Patrick barked, and the Truce Hall was set in motion.

The concrete bunkers on the hills overlooking the beach were flooded with activity. Heavy machine guns were already prepped and loaded for the coming conflict. Several miles away, artillery batteries opened fire. The fire support was being led by spotter teams in the bunkers.

With a high-pitched whistle, artillery rounds screamed through the air. As they splashed down in the lake and detonated, plumes of water were blasted high into the air. The water crashed down on the military transports laden with Biogtech soldiers in a glorious cascade. The mindless robots were undaunted by the violent attack, some of them laughing in a dry, robotic tone. Armed with a host of automatic weapons, the Biogtech soldiers were ready to destroy anything in their path.

The artillery attack was relentless. Zeroing in on several

transports, the artillery rounds struck true, blasting two of the transports before they reached the beach. As the explosions tore the vessels asunder, water rushed inside the hull and the transports sunk almost immediately. Steering around the destroyed wreckage, the remaining troop transports sped forward with the intent of releasing their deadly cargo of soldiers on the beachhead. Bobbing over the fierce waves, the boats came in fast, leaving the defenders of Dakota Beach stunned.

The remainder of the transports hit the beach. Dropping their gangplanks, the ships rapidly unloaded a horde of Biogtech soldiers. Cackling, the mindless robots prepared to slaughter the defenders without mercy. Shambling forward, the army of mechanical soldiers scanned for enemies.

The concrete bunkers roared to life. Hundreds of high caliber rounds tore into the tightly ranked robotic soldiers moving forward. Piercing several ranks at a time, the machine gun bullets were shredding the robotic death machines without recourse. The Biogtechs fell like dominos under the harsh assault. Hydraulic fluid was pooling in the hulls of the boats as they collapsed. Limbs and shreds of plastic and metal were blown free as the harsh gunfire demolished the robotic death machines.

Though the full force of the Iron Kai beach defenses was being unleashed, the enemy was too large in number to be fully repelled. Gaining a shaky foothold on the beach, the Biogtechs returned fire. Hundreds of rounds charged forward, slamming into the fortified bunkers. Chips of concrete were blasted free as Iron Kai troops took cover. The return fire was so aggressive that many defenders were caught in the hail of gunfire, falling lifeless to the ground. Using the advantage gained by pinning many members of the machine gun bunkers, the Reaper Kai forces took the opportunity to unload the remaining ships. Several hundred Biogtechs had survived the lake crossing and were firing on the defense positions.

Screams rose on the hillside. "Mortar teams! Fire at will!" a veteran officer ordered. *Thump. Thump. Thump.* The small tubes popped as the weapons were ejected from their firing chambers. The explosive projectiles arced in the air and slammed into the beach with devastating results.

The weapons boomed out as they detonated. Sending

shrapnel into the robotic soldiers, the mortar rounds battered the densely packed troops, shredding them in a flurry of explosive force. Staggering in the confusion, the death machines were unable to retaliate. The staunch defense of the Iron Kai beach was formidable. Many of the soldiers defending the city of Stonen had seen years of conflict and weren't about to yield the beachhead to enemy forces.

Breaking through the mortar fire but taking heavy losses, the Biogtechs moved on without a hint of self preservation. As they fired back, several machine gunners were struck with submachine gun rounds which felled them inside their concrete fortresses. For each gunner that fell, an eager soldier rallied to the cause and took charge of one of the powerful machine guns.

Having moved about twenty yards up the beach, the robotic soldiers needed to move another thirty to reach the bunkers. The Iron Kai, tracking their advance, fought harder, not wanting to give another inch of ground to the mindless servants of the Reaper Kai.

Hot white tracer rounds were streaking forward, granting the gunners a striking advantage. The machine gun nests were so closely positioned to one another that adjacent bunkers often overlapped in their effective firing arc. With dense gunfire shredding the Biogtechs and mortar rounds still slamming into the tight formations, the enemy offensive was a futile attack. It took the Iron Kai defenders a mere ten minutes of heavy fighting to finally kill the last of the Biogtech soldiers. The attack had been driven back, and the Iron Kai forces were victorious.

As he monitored the end of the sparse, almost feeble enemy offensive from the Truce Hall, Emperor Gunther's expression was grim. The cogs and gears were grinding away in his head. Cheers of success were echoing boisterously behind him. The first Reaper Kai offensive against the capital city of Stonen had been repelled. Throughout all of the cheering, no one but Field Marshal Patrick had noticed that Gunther hadn't joined the general jubilation; instead, he was lost in thought.

"Silence!" Patrick boomed, and an eerie quiet fell over the Truce Hall. The military advisors, seeing the concern on Gunther's face, were suddenly filled with a sense of confusion. Moving forward, they surrounded their leader and waited for him to speak.

Rubbing his beard, Gunther turned to one of the radio operators. He had a suspicion about the attack, and was seeking evidence to prove his theory.

"Transmit a message to the troops near the beach. Send out several squads and search the Biogtechs."

The radio operator relayed the order and three squads moved down to the beach, searching the remains of the Biogtech soldiers.

"What is it?" Patrick quizzed Gunther with a concerned look on his face.

"It's just a hunch. I'm betting that none of the Biogtechs had any extra ammunition on them," Gunther replied, inspiring a general confusion in the room.

"Extra ammunition?"

"Yeah, it's just a hunch, but I think it's sound."

The radio crackled to life in the Truce Hall. The soldiers on the beach had found an odd fact about the enemy troops. Gunther's hunch had been correct.

"Sir, none of the Biogtechs had any spare ammunition. We searched several of the robots and they didn't have even one spare magazine of ammunition among all of them," the soldier on the beach informed the military advisors in the Truce Hall.

Shocked by the revelation, everyone stared at Gunther in disbelief. Only a minute ago, their Emperor had seemed to simply divine the eerie truth about the Biogtech attack.

"How the hell did you know?" Patrick almost stuttered, looking at Gunther in amazement.

"This battle was not about gaining ground, taking territory. This battle was about mathematics." Stroking his beard as he spoke, he looked at each of his advisors in turn.

"Mathematics?" an advisor echoed in confusion.

"Yes, mathematics. The enemy marched twenty yards up the beach. If they had moved another thirty yards, they could have overrun the machine gun nests. This battle was not about taking the beachhead. The Reaper Kai probed our defenses. Why give your troops extra ammunition if you expect them to be destroyed? Why send only five hundred troops when you have thousands in reserve?"

The advisors were stunned and did not respond. Seeing that none of his subjects had a clue, Gunther sighed and answered his own question.

"I will tell you why," Gunther addressed his mystified audience. "The battle was used to calculate how many troops it will take to overrun our defenses."

"That's absurd. Why would the Reaper Kai sacrifice five hundred soldiers to gauge our defenses? It doesn't make any sense. The loss of resources is too drastic." Field Marshal Patrick was the only one confident enough to counter his theory.

"Your point is well put," Gunther conceded, but something else was unsettling him. "We know that the enemy did not outfit their troops to win this fight. That fact is indisputable. That leaves us with a horrifying prospect. If the Reaper Kai did use these troops to probe our defenses, we have vastly underestimated their troop production. We have seen it before. Remember Rust Spire? Remember Markov? In both instances, the enemy managed to bring a vast number of enemy troops into battle without us even knowing where they came from. The number of enemy troops is much larger than we originally estimated, and it seems to be growing. The five hundred troops laid to rest today were nothing to the Reaper Kai. If they can waste that many troops to probe our defenses, the enemy is undoubtedly preparing for the complete destruction of this city. They now know what they need to overrun our defenses. We must either destroy the Reaper Kai or dread the day they come knocking once more."

The speech made sense. The battle which had just ended hadn't been about gaining territory; it had been about mathematics. After enemy leaders surveyed the results of the attack, a time schedule would be put in place. When the Reaper Kai had enough troops, the Iron Kai would be crushed and the war would be over. The forces of evil would finally dominate the Darken Realm and build a dystopia of evil.

"The prospect of enemy troops overrunning the wasteland and the remainder of the Darken Realm has to be dealt with. With the heavy number of enemy troops, I seriously doubt we have located all of the Biogtech production facilities. I want scouts dispatched to seek out any clue or rumor of enemy troop production. Hopefully, if an enemy production site is out there, our scouts will find it," Gunther added.

His theory of Biogtech production facilities outside the

enemy capital was met with intense skepticism.

"There are no such facilities, my lord. We simply underestimated the original troop numbers before our siege of the enemy capital," a weasely advisor scoffed the emperor's assertion.

"No such facilities? We have cut off all supply routes out of the enemy capital for many months now. If the troops are not coming out of the Reaper Kai capital, they have to be coming from someplace else. We did not underestimate original troop numbers! The enemy has additional production sites not yet located," Gunther shouted at his defiant advisor, who fell silent, conceding to his fury.

As their leader's rationale gradually sunk in, it began to make sense. The elusive Reaper Kai had many secrets, including quite a few which still needed to be unraveled. With this grim reality setting in, everyone in the Truce Hall began to wonder when the armies of darkness would return once again to test their defenses. Time was running out for the Iron Kai and their allies.

"What are we going to do?" Patrick questioned as a shiver of terror rolled down his spine.

Staring back with a solemn look, Gunther replied, "We are going to give the Reaper Kai hell, absolute hell. Our scouts need to locate any other Reaper Kai production sites and our army needs to destroy these sites. After that, we need to pray that Nova 7 is still out there, fighting through the ruins, searching for a way to bring about nuclear retribution before the forces of the enemy overrun our defenses."

"Nova 7 is long shot, my lord," an advisor responded in an exasperated tone.

"Long shot?" Gunther grumbled. "It's not a long shot, it's now our only chance for survival. Though we are keeping the enemy capital city under control for the time being, we are running out of soldiers and will eventually fail. We must trust in Banion O'Neil and his team if we are to survive this conflict."

Growing weary of his council, Gunther turned his back to them and walked over to the mighty windows overlooking the lake below. With a calm breath, he surveyed the carnage below. Dead soldiers were being collected and carried away on stretchers. Seeing their lifeless bodies upon the beach struck him as an ill omen. The war was once a distant menace, relegated to chaotic radio transmissions and vague uncertainty. Now, evidence of hate was littered all over the beach below. Shaking his head, Gunther secretly began to wonder if he was witnessing the last days of his empire. Hiding his fear, he resolved to never yield to such a dark vision.

Chapter 36
The Road to Scarskin

Sister Nightshade was on horseback. Her mount twitched nervously, rearing back, shying away from the disturbed patch of sand. Two Reaper Kai priests were beside the matron of evil, grappling with a captive who was bound and gagged.

The captive maiden did not know what lay ahead, but she trembled nonetheless. Slaves and captives had been disappearing every few days, never to be seen again. Always they disappeared at night, and piercing screams echoed in the darkness just after they were led away from camp. Whatever lurked in the shadows of the clearing was a thing born from tortured dreams, a haunted beast born of darkness, and a mystery the young woman was not prepared to discover.

Nightshade looked up into the heavens. Darkness had washed over the desert, bathing it with inky blackness. Near the horizon, the moon was rising, full and crisp. As she gazed at the full moon bathing the wasteland, the evil Reaper Kai priestess knew that what lingered in the clearing was none too fond of the light. "Take her to the clearing and step away."

The Reaper Kai priests, garbed in crimson red, dragged the frightened woman into the clearing. Thrashing back and forth, the captive tried desperately to escape. Her fight was in vain, and the two priests overpowered her quickly, lifting her wriggling body off the ground and moving her toward the patch of disturbed sand. Leaving her near the center of the clearing, the two priests made haste to escape the area. They passed by Nightshade quickly, determined not to witness what was about to happen, guided primarily by their fear. As they passed Nightshade, she dismounted and they led her trembling horse away from the scene.

Knowing that she was alone with the captive in the clearing, Nightshade eyed the prisoner with a stern look. The captive looked back and beheld Nightshade's horrific visage. Pus and boils covered her face. Lesions blistering with blood were oozing down her face. Nightshade's tortured eyes were so heartless, so cold, so dark and filled with sickening malice. Beginning to tremble violently, the captive felt the constriction of her bonds once more, and tried to wiggle across the sand in a feeble attempt to escape.

"Don't struggle," Nightshade cooed serenely, without a hint of emotion. "You should be honored. Soon, you shall be immortal." With a sinister smile, the matron of evil stared at her intently.

A tremor of terror rocked the woman. Shaking violently, with tears streaming down her face, the captive whimpered. Her sobs of terror did not go unnoticed. A creature, a wretched monster born from the underworld, was listening intently to the whimpering woman, elated by the evidence of her fear. Suddenly, it hungered. Giving in to savage instinct, the demon sought to liberate itself from its hiding spot beneath the sand.

A quake rocked the earth. The sand shifted quickly as it heaved. Something was pushing itself to the surface, trying to emerge from the earth itself.

Still upon the sand, the woman felt the vibrations below her. Squirming in fright, she tugged at her bonds. As she wrenched her arms back and forth, the rope dug into her wrists, tearing the skin with a stinging pain. Her whimper had been replaced by a primitive paralysis of helplessness. The tears erupted from her eyes as she struggled to breathe. Wheezing and trembling, the captive was on the verge of losing consciousness, her terrified mind wanting to simply surrender in an attempt to evade the terror.

Erupting from the sand, an enormous green arm surfaced abruptly. Another tremor preceded the appearance of a second arm. Between the two arms, a sinister head emerged from its hiding place. Bright yellow, piercing eyes looked forth as sand trickled off the beast's enormous head. With a spasm, the creature wriggled and began to liberate itself from the earth. As its body emerged, the voices of the damned rose in the darkness. Tormented screams tore at the night air as hundreds of faces upon the beast's body wailed in terror. Each face had once been a victim. Each victim had become permanently trapped inside the body of the beast, their souls bound within the flesh of the fearsome demon.

The full moon was now shining brightly upon the desert landscape, bathing the sand in a warm, holy glow, and the beast squinted at it, roaring with displeasure. Holding its mighty hand to block the light, the Abomination was dismayed by the sickening moonlight. The demon was born from pure evil, and the sun was its bane. Normally, the darkness was devoid of light, but when the moon was full, the light reflected from the sun hurt the demon, making it weaker, making it want to hide from the light. Blinking and roaring, it searched the surroundings hastily for its meal.

Having grown considerably over the past months and now standing almost three stories tall, the Abomination hobbled forward. Its right leg had been mangled in a past battle and still had not fully healed. The injured leg was bowed, looking almost as if it would collapse under the enormous weight of the beast. Though crippled, the beast still had a mighty hunger. It wanted warm flesh. It needed a tasty meal, a soft body to arouse its palate. Feeling wild hunger overwhelm its senses, the demon sought a new victim to feed upon. Still squinting in the light, it struggled to locate its prey, bright moonlight burning its eyes.

So fearsome was the beast that the captive woman was too terrified to look upon it, desperately averting her gaze from the monster. The demonic magic bound into the creature made it impossible for the meek to look upon it. As she fought to escape, the unknown began to play at her mind. What was the fiend before her? How could such a monstrosity walk the earth? Gripped with confusion, her overactive imagination fed by fear, she experienced a fresh surge of terror as she wriggled in her bindings, screaming in panic.

Her cries captured the attention of the Abomination. Sniffing the air, it roared, "I hunger! It has been many days since my last meal. I will savor the taste of your flesh as you slide into my gullet!"

Screaming again, the woman struggled like a fish out of water. Coming before her, the beast gripped the screaming woman in one hand. Bringing her to its face, it smelled her. The stench of death erupted from the creature as its open mouth salivated wildly, hungering immensely. As the monster smelled her, the faces upon it stared at the captive with a mixture of pity and anger. The hundreds of trapped souls bound to its body began to whisper suddenly, begging for mercy, begging for their suffering to end.

"Save us!" one of the human faces groaned, its eyes looking out from the stomach of the Abomination.

"Help me!" another face moaned, its eyes shut but its mouth moving back and forth on the creature's chest, speaking in soft, measured bursts of madness.

The horrifying occurrences were paralyzing the captive woman, who was now only inches from the beast's mouth. Darting out, the slime-coated tongue of the Abomination sampled its next meal, brushing across her face as she flailed at the beast in vain. Finding her taste pleasing, the Abomination tipped his head back and jammed the flailing captive into his mouth. The demon constricted its throat, squeezing the woman down into its stomach and swallowing her whole. Wriggling all the way down, the captive descended into the beast's stomach, then fell still. Her struggling had stopped. Groaning, the Abomination grew slightly larger as the life force of his newest meal was added to the creature's body.

A boil appeared on the creature's back. Swelling quickly, it burst open, pus and blood dripping out of the pulsating wound. As the boil ceased to writhe, a new face, the face of the woman just consumed by the Abomination, appeared. With a scream, she looked around, finding herself permanently bound into its body. Shaking her head back and forth, the woman was filled with terror. She wailed in fright, unable to comprehend that she was now part of the Abomination's flesh. Looking around, she caught sight of other souls and victims embedded in the monster's flesh. They all looked back with frantic, haunted eyes. Panic drove the woman, a feral fright that would never leave her. Sobbing in terror, she looked on as the other victims babbled, already driven mad by their imprisonment in the creature's flesh. The horror of her reality began to set in. She could never be at peace until the Abomination was slain. Nightshade was correct; the captive was now immortal, trapped inside the flesh of the creature forever.

After feeding, the demon felt at ease, for the moment. Its

hunger had been temporarily sated. Seeing Nightshade in the clearing, it moved over to consort with her.

Only souls of pure evil or pure virtue could look upon the Abomination. All others were unable to view the creature. Nightshade was such a being, having given herself completely to evil. She looked upon the demon with dispassion.

"I grow tired of slave flesh! I hunger for something more substantial! Perhaps a noble... perhaps something younger..." the demon growled, hiding its face from the moonlight, feeling the reflected sunlight weaken its body and resolve.

"Maybe in the next village," Nightshade replied.

"The next village? Where is this wretched Scarskin? I grow tired of hiding in the sand during the daylight so we can hunt out and destroy this puny little village." The beast was referring to its casket, which had started out as an enormous box pulled by forty oxen. Over the months, the Abomination had grown too large to travel inside the wooden crate. Instead, it would bury itself in the daytime, hiding from sunlight, which could turn its body to burning ash. At night, the demon would emerge and the army would travel until the rays of the sun shone near the horizon.

"This village of Scarskin holds enemies of great importance. One of the villagers is immune to our power. The brat that attacked Vertigo in Rasheed is now whispered to slay Reaper Kai without suffering a single wound or injury. This Jared of Scarskin is now a legend of the wasteland, a warrior who is immune to our demonic power. Father Vertigo believes that this boy is a descendent of the lost tribe of Ceibla Moralis, the betrayers of our race who turned from the darkness to consort with angels. The descendants of Ceibla Moralis are undoubtedly psychic as well. No psychic can harm another psychic with their powers. We must seek out this village and destroy them before they can muster a force to rival our own. Without the advantage of our psychic powers, our army is weak." Nightshade herself had been convinced by the rumor that Jared of Scarskin was a descendant of the lost tribe, and was oblivious to the true source of Jared's power.

"The descendants of Ceibla Moralis will not survive the war. I will feast upon these traitors and bind their souls to my body. I will enjoy hearing their screams and pleas, bound inside my flesh for an eternity," the abomination roared in response to Nightshade's theory, anticipating the destruction of the lost order.

"Our duty is to find this lost tribe and slay them all. Finding this tribal village is of utmost importance."

"Indeed! When this village is located, nothing will hinder our victory!" the Abomination boomed. It squinted once more in displeasure at the bright, full moon. "I grow weary of this sickening moonlight. I feel weak and wasted in this dull glow from the full moon. It is time for me to rest once more." With that, the evil demon began to burrow down into the sand, hiding itself from the light of the full moon. As it submerged itself, the screams of the countless souls bound to its body grew silent, their pleas deafened as the Abomination burrowed deep into the earth. Just as quickly as the demon had emerged, it disappeared from sight once more. The disturbed patch of sand was once again quiet as the enormous demon hibernated, digesting its newest meal.

As silence filled her ears, the matron of the Reaper Kai was left to ponder the situation alone. Nightshade looked at the disturbed sand and then toward the moon. An ominous feeling overcame her as she gazed at the bright moonlight. A tremor of indecision flowed through her. *Something* unsettling, a minor disturbance in her usual confidence, hovered at the edge of her consciousness; a wavering fear wracked her soul. Trying to focus on the feeling, Nightshade attempted to discover the source of her concern, to no avail. Shaking off the eerie sensation, Nightshade headed back towards the Biogtech encampment. As she walked, the evil priestess thought about her objective: the destruction of all the inhabitants of Scarskin. When she found the village, there would be no mercy or compassion, only reckless slaughter.

Smiling in the darkness, Nightshade reached the camp and surveyed her army. A tremor of joy filled her as she viewed her robotic soldiers and loyal Goat Minions. War was near, and she was anxious to carry out a sinister act of hate and murder in the tribal village of Scarskin.

Unfortunately for the inhabitants of Scarskin, none of Ceibla's order was in fact in their bloodline. The hapless desert inhabitants would be subjected to a surprise attack against which they would be hard pressed to defend themselves, and which might result in devastating consequences.

Chapter 37
The Oracle of Heaven

A blinding white light filled his consciousness. Wanting to flee, the man tried to run, but fell to the ground instead, his legs ceasing to move or to function. With terror filling him, he tried to scream, but his voice was silent. The man was helpless, thrashing on the ground as the swirling white mass of light moved forward with frightening speed. As he closed his eyes, the bright sphere of light exploded with a blinding flash.

After the immense light show had dissipated, the man opened his eyes and gazed upon the strange sphere. Flashing brightly, it rotated at a breathtaking speed, spinning downward until it was shaped like a disk. The man cowered, feeling powerless before the event unfolding in front of his eyes. Finally, the disk stopped rotating and a shimmering form emerged from within it.

Although shaped somewhat like a human, the glowing entity had no real features, only a vague shape. It moved forward and stood before the fallen man, declaring in a booming voice, "It is time."

Feeling terror fill him, the man tried to dispel the vision, the same vision that had been plaguing him for well over a year. The glowing entity would visit the man in his dreams on an almost nightly basis, speaking of war and distant battles. Such an ordeal had made the man's mind fragile. In an attempt to escape the entity, he tried to ignore it.

"Your ancestors have grown decadent. The Reaper Kai have now destroyed almost all bastions of truth and justice. It is time for you to make your pilgrimage to *her.* "

The spirit was referring to a strange woman with dark skin and glowing, brilliant blue eyes. She was kind and wise, but tortured by the very will of the devil. The man dreamed of *her* many times. He sensed she was some sort of beacon, a focal point for something profound.

"I do not want any part of this war," the man whimpered, still trying to flee from the entity.

"The war will purge the land of all that is good if you do not concede to your responsibility. You are one of the descendants of Ceibla Moralis. Your heart is pure and your soul is clean. It is because of your deep respect for justice and love that you have been chosen to perform this pilgrimage." The glowing entity spoke, but no sound came from its crude mouth. Instead, the words were forced directly into the man's mind.

"I have followed the teachings of Ceibla, as have our entire tribe since his death. I fail to see how one man can make a difference in the war engulfing this land," the man protested, hoping that the spirit would take pity upon him and simply leave him alone.

"There are others, like you who are to travel to *her.* Do not fret. A series of events has begun which, if seen through to the end, will spell the ruin for evil. If your faith is strong and you believe in the righteous path, you shall not fall in battle."

"In battle?" The man was horrified. "I am no soldier."

"You will be called to battle nonetheless."

A moment of silence ensued. The man wanted to fight against the spirit but he knew he could not. Finding courage within himself, he nodded in agreement to the entity before him. "What am I to do?"

"You are to leave tonight. Do not tarry. Ride into the darkness and follow the setting moon."

"Tonight? But what of my family? What of my wife and children?"

"To truly save them from the coming darkness, you must do as I say."

"Where do I go after the setting of the moon?" The man was baffled by the instructions from the glowing entity.

"Your dreams will lead you." With a whisper, the glowing form began to dissipate. Within a few seconds, the entity was gone.

Daniel awoke from his slumber with a gasp and sat straight up in bed. He was covered in sweat and breathing heavily. His wife, Rebecca, had awoken earlier and was staring at him with a look of concern. Her husband had been muttering in his sleep while thrashing around, a thing he had done frequently in the past few months.

"Are you all right?" she quizzed him, taking his arm and massaging it with a firm grasp.

Gripping his head in his hands, he shook it wildly, wiping the sweat from his brow. Daniel was noticeably alarmed, rattled by his strange dream. Seeing him so lost in thought, his wife was rapidly becoming terrified.

"Daniel?" she said in a quick, frantic tone. He did not respond. Growing more anxious, Rebecca grabbed him again and shook him briskly. "Daniel!"

He still did not respond. Fear had taken his soul toward a dark place, in which nothing seemed certain. Was the dream real? Was it something to be ignored? The internal battle within him was mounting. Was he simply crazy or did the spirit world have a plan for him, some sort of grand destiny?

Fearing that her husband had gone utterly mad, Rebecca tried to snap him out of the malaise that had gripped him. He did not answer her, and instead began to weep, tears rolling down his face. Gripping his face with his hands, he sobbed uncontrollably.

"Daniel, what's wrong?"

No response.

"Why won't you answer me?" she asked, agitated.

"I have to go now…" He spoke with a distant look upon his face.

"Go? What are you talking about? Where are you going?"

"I love you very much," he said, standing up beside their bed. Turning from her, he started to walk away. Rebecca grasped his hands tightly, trying to force him to stay by pulling him back to bed. Panic was driving her actions and her mind was spinning frantically. She did not know what was wrong with her husband, and fear was gripping her soul. No matter how hard she tugged on his hands, trying to pull him back, he resisted. Finally, conceding defeat, her grip loosened and he pulled away from her.

Putting on a pair of pants and a shirt, he walked into their kitchen and grabbed a loaf of bread, then threw a brick of cheese and a lump of jerky onto the kitchen table. Rummaging through the cabinet, he located a pair of canteens. He filled them quickly as his wife looked on in horror, not understanding what was happening. Grabbing a burlap sack, he placed the provisions in the bag and pulled on his riding boots.

"What are you doing?" she sobbed. Mucus was dripping from her nose, tears streaming down her face. "Don't leave me!"

"I must. I will return."

"Please don't leave me," Rebecca pleaded again. Forcing his own emotions from his mind, Daniel knew that he had to fulfill his duty. "What did I do wrong? Did I do something to hurt you?"

"It's not you. I love you very much. That's why I have to leave."

"I don't understand," she sobbed.

Daniel stopped dead in his tracks. He breathed in and let out a heavy sigh. "I don't understand either. I am sorry. I have to leave."

Grabbing a sleeping bag and some blankets, he was ready. Fighting back his own tears, he moved out of their house into the dark night. A blast of cold hit his skin as he stood outside the house. Still sobbing, Rebecca stood in the doorway, wishing she knew the reason for his strange behavior.

Turning around, he looked at his wife. With lower lip quivering, he ran over to her. He gripped her tightly, hugging her with all of his might till he lifted her off the ground, holding her tight against his chest. She welcomed the embrace, secretly wishing with all her heart that the power of her love could stay his madness.

Bringing his lips to hers, Daniel kissed his wife. As their lips merged, they could taste their own tears, streaming down their faces. Then he pulled away. Wiping the tears from his face, he spoke in a whisper. "I am so sorry. I love you so much."

Turning away, Daniel untethered his horse. He climbed up, secured his belongings, and dug his heels into his horse with a forceful kick. The horse reared back with a loud neigh of alarm, then bolted off down the dirt road, sprinting into the night.

Daniel never looked back.

In the dim light, Rebecca made a feeble attempt to follow.

Running along the road, she chased him as his form disappeared into the shadows of the night. "I love you!" she screamed. "Do you hear me? I love you!"

In the blink of an eye, he was gone; her beloved husband had vanished into the darkness. Falling to her knees, she sobbed until the tears running down her cheeks froze in the cool night air.

Rebecca knelt upon the road, staring into the blackness. The world had grown quiet. Feeling defeated, she wondered if she would ever see him again. In the dim light, she closed her eyes and let the sound of her own heart fill her ears. As it beat in her chest, her pulse grew softer as the trauma began to wane. Wiping the tears away, she made the long walk to the front door of their silent house. Closing the door, she moved to the children's room. Seeing them sleep peacefully filled her with hope.

With the fatigue of the painful event guiding her, she collapsed into bed. Her hand stretched out and felt the spot where Daniel usually slept. His side of the bed was still warm. In the dark silence, she felt the warmth of his body fade. Her hand caressed the warm spot as memories flashed through her mind. When her hand grew cold, she drifted off into a deep slumber, a lonely dream world filled with memories of her troubled husband.

Chapter 38
Taming the Monster

"I should have seen this." With tears streaming down her face, she shook her head back and forth in frustration. Alone at the edge of the ridgeline, Mineera stood and let sorrow take her.

A cool wind was blowing in from the western reaches. The sky was set aflame with violent orange hues, dispersed amongst somber purples. As the wonder of the coming darkness filled her with a lonely misery, she stood and stared down at a lonely grave upon the top of the hill.

Brilliant green grass danced in the wind and she smiled softly, feeling utterly defeated. "I am sorry I failed you," she murmured in a mournful tone. A flood of tears erupted from the corners of her blue eyes. The pristine liquid rolled down her dark skinned face. Though she tried to fight back the emotion, she was lost. Her right hand stretched out, shaking violently. Gazing intently at the grave, Mineera imagined she was touching it with her hand.

She found herself collapsing, kneeling down before the soft mound of earth. Placing her hands in the soil, she began to sob. She gripped the earth in both fists, pounding at the grave. "I am so sorry I failed you..."

Thrashing about like a fish, Mineera was still caught in her slumber, trapped in the disturbing dream which had invaded it. As she tried to fight the strange vision warping her sleeping mind, the psychic was feeling dismayed on many levels, and with good reason.

Something had crept from the darkness, hearing her cries. A shadowy figure approached the slumbering Mineera and knelt down, hunched over her. Her acute senses screaming a warning, the woman opened her eyes and felt a presence very near her. She threw her arms upwards and felt them strike someone in the darkness. Filled with fright, Mineera was about to screech out uncontrollably.

"Hold on!" Banion said in a soft tone, grabbing her arms.

The fear abated instantly at the sound of his voice. With a sigh of relief, she caught sight of Banion kneeling before her, touching her arms softly. Blinking several times, she stared into his harsh, dark eyes with her blue ones. For a brief moment, they gazed at each other with quizzing looks. Whereas once they had been the fiercest of enemies, a bond had recently formed between them. Their deep mutual mistrust had been replaced by a loyal respect.

He smiled at her and rubbed her forearms gently. "You were thrashing around again," Banion said in a matter-of-fact tone.

Mineera was stunned. Almost every night, she had horrid dreams and he never seemed to care. This night, however, the gunfighter had seen fit to ease her torment by waking her. It was a kind gesture, which had caught Mineera off guard. Smiling back at him in the darkness, she sat up and eyed him with a soft gaze.

"I didn't want you to have bad dreams again. I think you barely get enough sleep as it is." Moving back a little, Banion sat down, his eyes remaining fixed upon her face.

Still somewhat stunned by his kindness, Mineera required a brief moment to collect her thoughts and formulate a suitable response. "I think this is the first time you have ever woken me up. Most of the time you seem..." She was about to say something brutally honest, but her sensibilities got the better of her.

"Rude and uncaring?" Banion responded with a chuckle.

"Well... yeah..." she agreed hesitantly.

Amused by her frankness, he smiled again and stared at her with an unwavering gaze. Where once his eyes had been fierce and filled with hateful anger, an almost jovial happiness now gripped his features. Stretching out with her feelings, she sensed that his previous aggressive wall of malcontent was now a lighthearted calm. Banion's emotional state had changed, at least for the moment. It was the first time since she had met him that he had let his guard down completely. Not knowing what to think, she was curious about his thoughts and pressed onward with inquisitive conversation.

"Why did you wake me?"

Looking away from her, Banion surveyed their surroundings. They were still on the shoreline, at a small campsite, resting from their long journey into the northlands. Since the wretched Lavosi had joined them and agreed to lead them into the north, Nova 7, with an extra member in tow, had been pressing toward the military installation that the vermin king had claimed held the prize of all prizes, a fully functional nuclear warhead. Jolted back to the conversation, Banion ignored their surroundings and focused on Mineera again. "I woke you because I have been thinking a lot about you lately."

"Thinking about *me*?" She was intrigued again, her interest piqued. "I don't get it."

Sighing, he rubbed the scar on his face with his right hand. The irregular flesh was a knotted mass of tortured memories. Shaking his head in wonder, he chuckled again, then replied, "I can still see you, standing upon that ruined column of concrete, staring down the entire world. Every time I close my eyes, I think of that day you put your life at risk to stop the war between the Slumlanders and the rat tribe. Without a single shred of self preservation, you risked your life to stop a war that was not your own." Shaking his head again, he reiterated Mineera's seemingly incomprehensible actions. "You would have given your life for complete strangers."

Looking back at him, she smiled with a sense of empowerment. "Yes, I would have sacrificed myself that day. I have seen and caused so much bloodshed in my life, it was the least I could do. Letting myself perish for the hope of peace would have been well worth the cost."

"I have been thinking a lot about what you did. Before meeting you, I would have never thought such deeds were possible in this horrible world. Your act of devotion to peace has made me think about my own life. Once, I thought of self sacrifice as foolish, but now I see it as a noble cause. Your example of selflessness has inspired me," he said earnestly.

Mineera blinked several times, thinking it was some sort of ruse. Had the crazed Banion just complimented her? Probing his emotions, she found his words to be genuine. The reckless leader of Nova 7 was proud of the former Reaper Kai. It was a profound moment in Mineera's life. For the first time, she was no longer an outsider. Mineera had gained the favor of their dour leader. Finally feeling as if she belonged, Mineera's heart leapt.

"I don't know what to say." Her tone conveyed both how flattered she was, and her new sense of empowerment.

"You don't have to say anything." Banion smiled. "I am glad that you have changed my perception of the world. I now feel that selflessness, not hatred, will win this war. It has been a lesson that I sorely needed to learn. I am glad that my view of things has changed for the better."

Returning his gaze, Mineera smiled at him in the dim light, her eyes glittering in the darkness. "You're welcome."

With a sigh, Banion became restless. Growing tired of such sentiment, he rose to his feet. He grabbed his assault rifle, resuming his stern gaze as his normal, edgy self emerged once more. The momentary feeling of compassion had ended. Aggressive, combative feelings returned to him as he stared at her with a look of dispassion. He was all business once again. "I'm gonna walk the perimeter," he said in a brisk tone.

Banion merged into the darkness of the night, disappearing from sight.

Mineera sat alone and thought about the exchange. In a brief fraction of time, the hate-filled monster known as Banion had been able to forget his tortured past. In the blink of an eye, a man driven to insanity by acts of hatred had learned to love again. If a reckless man such as Banion could learn peace, there was still hope for the world. Smiling in the darkness, Mineera sighed and pulled her sleeping bag around her. The world had made a slight move towards peace. Banion turning his back on hatred and violence was but a single step, but it was a step nonetheless.

Chapter 39
The Outsider

Anchored securely in place and resting upon the waves of the open ocean, the ship was rocking back and forth, bobbing up and down. Still on their journey northwards, Nova 7 had stopped once more for the night. The coastline was a radioactive hell hole. All of the cities in the region had been ravaged by nuclear fire during the apocalypse, and their blasted rubble was still radioactive, making it impossible to moor the boat and find sanctuary, or locate a suitable campsite. Seeking an alternative, the team decided to rest on the ship overnight, upon the ocean waves.

The ship was silent except for a haunting whimper. Once again, the dream world was waging a war against the beleaguered psychic. Visions and intense dreams had taken control of Mineera's slumber once more. As usual, she was thrashing about like a fish inside her sleeping bag, arms flailing around as she spoke in fragmented sentences. Trying to fight the strange vision warping her mind as she slept, Mineera was feeling disturbed on many levels.

During her wild antics, Mineera had drawn the attention of something sinister. As dreams of darkness gripped her, a creature crawled from the night. Quivering in the darkness, a white form covered in ashen fur slunk forward. The creature sniffed the air, probing its surroundings for a scent, the scent of fear. As its nose twitched and its long whiskers flexed, the odor rushed into the creature's nostrils. Seeing Mineera in distress and helpless, the creature drew ever closer. With each step it took, its blood red eyes looked forward and a smile graced its lips. Silently it strode, ever closer to the slumbering woman, enthralled by her suffering.

Lavosi, the wretched mutant rat, was watching Mineera with interest. Her duress during the nighttime hours had recently become fascinating to the loathsome creature. Enjoying her suffering, the vermin monster had become accustomed to watching her sleep, often coming close enough to taste her fear.

Coming face to face with her, the albino rat sniffed her again, taking in the savory smell. Grinning, he looked her over with perverse, hungering eyes. The mutant got so close, his whiskers tickled her bare skin.

Sensing something was wrong, Mineera sat up, awaking abruptly from her slumber. Her sharp gesture caused her to collide with the strange creature brooding over her. With a flurry of movement, she pushed the form away from her.

Lavosi crashed backwards, landing on his coiled, pink tail. Grunting, he leapt forward and stared at her intently. Startled and breathing heavily as a result of her recent ordeal, Mineera simply looked back with a staunch, unwavering gaze. The rat king bobbed his head, uttering another defiant grunt. With a sneer, the loathsome creature disappeared into the darkness, his white form moving into the shadows like a ghost.

The past few weeks with the mutant had been tense at best. He was a tiring presence, constantly trying to wear them down and drive a wedge between them. Instead of yielding to his sinister presence, the team had grown closer, favoring each other over the wretched outsider that had joined them. Lavosi, though wicked, was completely transparent. Driven by jealousy and greed, the sinister mutant was easy to read. The days in which the team members squabbled endlessly amongst themselves had quickly ended. Instead of wasting their precious breath on bickering with one another, the members of Nova 7 had taken a silent pledge to stand together against their newest team member.

Their sense of defiance only made Lavosi bolder. The less they listened to him and the more they tried to ignore him, the more obnoxious and devious he became. Like an abused child looking for attention, his most recent antics were becoming more and more unsettling and downright creepy.

Banion had witnessed the entire event. The hammer on his revolver was pulled back, ready to slam into the primer on a fresh bullet. The gun sights were upon Lavosi the entire time, waiting for the dire second when he would feel justified in killing their sickening guide. Now that the loathsome creature had disappeared into the darkness toward the back of the boat, Banion eased the hammer on his revolver down with his thumb.

"I'm fine. He just startled me, that's all," Mineera called out to Banion. "Go ahead and get some sleep."

The wizened mercenary never truly slept. Tani had once said that their fearless leader always slept with one eye open, and there was great truth to that statement. Always ready for battle, Banion's stamina was a testament to an indomitable will. The wild mercenary had not slept more than twenty minutes at a time for the last fifteen years. This night would be no different.

"Go check on Tani. That damn rat headed toward the back of the boat. I've caught him trying to steal Tani's laptop computer the last few nights. Make sure he's safe, too." With a sigh, he tipped his hat over his face and attempted to 'sleep'.

Mineera gave a nod of agreement. Pulling her robes about her, she stood up in the rocking boat. Steadying herself upon her recently seasoned sea legs, the psychic took a few steps, moving toward the back of the ship. As she moved, she gripped the cold railing. Small waves broke against the hull as she headed toward the back of the ship. Hearing the wild ocean break across the boat was soothing to Mineera. Taking a brief moment of solitude to listen to the waves, she felt refreshed and ready to face off with the vile rat king once more. With a sigh, she continued her progress toward the back of the ship.

Finding Tani, she was elated to confirm that the mutant rat was nowhere to be seen. The scholar was sitting on the deck of the ship, a blanket draped around him. He had covered himself in such a way that only his head could be seen peering out of the blanket. As he worked diligently on his laptop computer, the scholar's eyeglasses were filled with bright colors reflecting off the computer screen. Seeing him in the middle of the night, engrossed in learning about the world, was refreshing. She smiled, reveling in his innocence and hunger for knowledge.

"How you doing?" she asked in a heartfelt greeting.

Turning his head, Tani smiled then turned back to his studies. Banging away at the keyboard, the tribal scholar continued his efforts. "I couldn't sleep. The boat still makes me a little sick."

"I know what you mean. I am still trying to get my sea legs."

"What are you doing up?" Tani asked, yawning deeply.

Not wanting to tell the tribal scholar the reason behind her restless night, she simply shrugged. "I couldn't sleep either."

Just then, a loud rasp broke the air. A groaning, gagging, nasal snort rang out. Jared, sleeping only a few feet away from Tani, was snoring the night away, mouth wide open, drool dripping down his chin. The warrior's arms were sprawled about him and his blanket chaotically draped around him. Jared's head was resting on the hard wooden deck of the boat. It was painful to behold, but the tribal didn't seem to mind; he was caught in a deep, snoring slumber.

Hearing the tribal shredding the quiet night with his snoring was comical. Trying not to laugh, Tani and Mineera snickered, attempting to keep quiet. Almost ready to lose it and burst out laughing, Mineera turned away from Jared and sat down on the deck of the ship with Tani.

Fighting off the cold night, she pulled her blue robes about her and looked over at the tribal's screen. Images of animals were framed in various windows on the screen. The tribal was using the encyclopedia contained within his ancient laptop computer.

"What are you doing?" she quizzed.

"I had an odd hunch. Something has been bothering me for a while now."

"Bothering you? Like what?" She seemed optimistic about her ability to help him discover some hidden truth.

"All of the animal species of the Darken Realm. I can't find a single reference or any evidence that some of the mutant species ever existed here on this continent before the Armageddon."

"You're right; many of them were not native to this continent before the fall of man." She was so matter-of-fact about the statement that it caught the scholar off guard. They had often debated the merits of science and religion, but had never had a conversation about biology. The confidence in her voice piqued the youth's interest.

"You know something about the animal species of the Darken Realm?" he quizzed, peering out from the warmth of his blanket draped over his shoulders and head.

"Of course I do." She smiled. "It's part of the reason I was branded a traitor."

Blinking several times, the scholar was hooked. Enthralled, he simply listened quietly and kept his silence, trying to encourage her to continue.

"Ever heard of a myth, a story about Godhome?"

Shaking his head, Tani indicated that he had never heard of such a thing.

Her expression solemn, Mineera began to recount a tale about the origins of the wondrous life forms inhabiting the Darken Realm. "It took a great many years for me to rise to power in the Reaper Kai order. After years and years of service, often involving brutal deeds of deception, deeds that I am ashamed of today, I was made head diplomat and became part of Father Vertigo's council. There I sat for a great many years, learning about the secrets of our dark order. One such secret was a legend as old as the Darken Realm itself."

Her gaze grew distant, as if some inner turmoil was gripping her. Recovering her focus, she began to speak once more. "In my travels as a diplomat, I was able to learn about many local customs and legends throughout the Darken Realm. I noticed immediately that almost every culture had legends and stories about a race of super beings, wise and powerful people who could shape life itself. Some legends called them Biotechnics, Gene-Crafters, or Genome-Mergers. I found that all of these names were convenient descriptions of the lost knowledge that this mysterious tribe possessed. Ancient texts speak of geneticists, and I became fascinated with the notion that there was a race of beings that could shape life itself. I brought this information to the head of the Reaper Kai order, and I was given the resources to investigate this myth. The intent was to find the source of this legend and to use the technology of the ancients to forge an army of genetically engineered super-soldiers." Shaking her head in disgust, Mineera seemed to be gripped by guilt. "I was so foolish during those days. I was filled with reckless pride and hateful ambition. All I craved was power and the ability to further the goals of our twisted order."

Mineera fell silent, gripped by inner turmoil. Seeing her distress and still wanting to know more, Tani placed his hand on her shoulder and rubbed it lightly. She responded to his reassurance and continued her story.

"I was given the task of uncovering the myth. Leading an entire team of Reaper Kai archeologists, we set out into the wastelands. Scouring the ruins, we quickly found entire civilizations that had died out hundreds of years ago. Each ruined village contained the remains of animal species not native to the Darken Realm before the Armageddon. Even more spectacular was the discovery that these animal species were not just ordinary animals, but had hands like humans, walked on two legs like humans, and were intelligent enough to build complex tools and buildings like humans."

"These animal species, they are all like Globulus?" Tani questioned, becoming increasingly excited. He had been perturbed by the mutant hippo ever since meeting him. His origin was mysterious and it seemed as if Mineera was about to uncover part of the mystery.

"You are correct. In our studies and excavations, we found dozens of such villages where mutant animal species had evolved. Though we could not decipher the languages of these lost tribes, we started to compile information about mutant animal tribes that still existed. In these studies, we found that every one of them had the same legend."

On the edge of his seat, Tani stared wide eyed at the storyteller who was holding his imagination captive. "What was the legend?"

"Do you ever wonder why all of the mutant animal species are shaped just like a human? Do you ever wonder why they all have hands and walk on two legs? You think blind luck and evolution had anything to do with it?"

Furrowing his brow, Tani pondered the quandary. "It doesn't make sense to me, it never has."

"Nor did it make sense to the Reaper Kai archaeologists. I think the reason that all these animal mutant species have the same traits is that evolution had nothing to do with it. I think all of these so called mutant animal species are not mutants at all. I think they were *engineered*."

"Engineered? By who?" Tani exclaimed.

"By the Progenitors!" A hiss rose from the darkness. Lavosi

had been hiding in the shadows, listening the entire time, eavesdropping on them. Slinking forward, the rat creature moved closer and stared at each of them in turn, his blood red eyes passing over them.

Eager to share his knowledge on the mysterious topic, Lavosi was jittery, almost ecstatic.

"Our glorious tribe has lived in the Concrete Barrens for hundreds of years. Our earliest ancestors, skilled in tunnel building, carved out a catacomb of passages beneath the ruined city. In the darkness of these tunnels, the first of our race scribed the origin of our tribe upon the walls so that none would ever forget our history!" Lavosi was shaking with excitement, his eyes bulging as he spoke. The display was so disconcerting that both Tani and Mineera edged away from him, putting extra distance between themselves and the hateful creature.

Undaunted by their discomfort, the sinister creature continued his tale. "The oldest part of the tomb tells of a place beyond the mountains where the Progenitors gave our glorious race life!"

"Gave you life?" Tani quizzed.

"Yes!" Lavosi hissed, his rat tongue quivering within his mouth as he spoke. "Knowing that the earth was battered and wounded, it is whispered in our legends that the Progenitors breathed life into many animal species, fusing them with humans!"

"Fusing them with humans? Like a genetic hybrid? Fusing human genes with animal genes?" The tribal scholar was horrified.

"Yes," Mineera replied. "The Reaper Kai also came to the same conclusion. The mutant animal species of the Darken Realm are genetic hybrids, not evolutionary mutations."

"So Globulus and his kind were created? For what purpose?"

"I believe the Progenitors made many hybrid species so that some could survive in this battered world!" Lavosi hissed, adding his own spin on the topic. "If the animals weren't adapted to the harsh, radioactive environment, none would have survived. Fusing animal genes with human genes gave the hybrids an enhanced ability to survive."

Shaking his head in amazement, Tani looked back at his computer. "It all makes sense. Many of the mutant... I mean *hybrid* animal species were not even native to this continent before the Armageddon. This is amazing." His mind spinning, the young scholar wanted more. "Where did all of these hybrids come from?"

"Godhome." Both Lavosi and Mineera spoke at the same exact moment.

"Godhome? The Reaper Kai were trying to find the laboratory that created all of these species?" Tani was horrified.

"Yes, and we were getting close to finding the ruins of Godhome. During this time, our order became reckless, using torture and bloodshed to push us closer to finding the sacred ruins. It was during this time that I began to question my faith in the Reaper Kai order. I think Vertigo could sense my wavering faith. I was taken off the project. During this time, I began to correspond in secret with the King of Rasheed. I promised him information about the Reaper Kai in return for safe passage and asylum. My last mission was to steal the data about Godhome. I failed and was captured for my treachery," Mineera concluded in a somber tone.

"Did you ever discover the location?" Tani asked excitedly. "Did the Reaper Kai find Godhome?"

"I am not sure. The last information that I found dealt with an unexplored section of the desert near a mountain range in the southern reaches, halfway between the Steel Crag Mining guild and the city of Rasheed. Apparently, Reaper Kai archaeologists had found *something,* but I am not exactly sure what. Whatever they found, it was a staunchly guarded secret. Only Vertigo and the high council knew what was discovered out in the barren wasteland. I was convicted of treason and sentenced to death before I was able to discover what the archaeologists uncovered in the desert."

"Godhome will never be discovered." Lavosi spoke in a soft tone, a distant look upon his face. "Such power could bring this fragile world to ruin."

"There are many powers that could bring this world to ruin..." Tani said softly, his words abruptly bringing about an end to the conversation.

"Indeed..." the rat hybrid replied, eyeing him with a piercing gaze.

Breaking eye contact, Tani retreated back into his own world. The conversation was over, but all three knew what was on their companions' minds. The ancients had been reckless, building technology that could barely be contained. The topic of uncontrolled power had struck a chord with each of them. In their own way, they were contributing to another atrocity for the sake of that reckless, ancient power. They weren't on a crusade to create life; they were on a crusade to end life in a blinding flash of nuclear fire.

The somber realization was an open wound for Tani and Mineera. It was a sore that would not heal. Retreating back into their own thoughts, the companions dispersed.

As the night wore on, Tani sat behind his computer and pondered the strange story.

"Godhome..." he whispered to himself in the darkness. The story was so fantastic, it played upon his wild sense of imagination. As Jared snored the night away, Tani pondered the possibilities of creating life itself. His mind spinning, the young genius was unable to sleep. Instead, he spent the remainder of the night thinking about the reckless power of the ancients.

Chapter 40
A Father's Pride

Bobbing over the waves, the ship pushed ever northward along the coastline of the continent. The once mighty cities, destroyed during the Armageddon, were now nothing more than nuclear craters, patches of earth where civilization had been reduced to mere ash and melted stone. Not a single portion of the coastline for more than a hundred miles could serve as a safe harbor. Instead, the companions were forced to stay on the ship the entire time.

Their goal was an old world oil rig known as the Steel Loft, about thirty miles off the coastline, which served as a way-station of sorts between the great northern forests and the southern coastline. The team, having moved slowly up the coastline, was only a day away from the Steel Loft. It would be the first time in weeks when they could leave the boat and refresh their dwindling supplies of food and fresh water.

Mineera was at the helm of the ship, steering it northward across the heaving waves of the ocean. The other team members were resting comfortably on the deck of the boat, spending a leisurely day in the bright sun. Banion was teaching the two tribals a mercenary battle tactic of some sort, using a series of hand gestures.

Both Tani and Jared sat attentively, listening intently as their eyes darted back and forth, watching the lesson unfold. Nodding in understanding, the tribals acknowledged that Banion's seasoned words were reaching them.

Taking a brief moment to watch them, Mineera smiled. The

tribals looked up to Banion as if he were their own father. They were eager for his attention, eager for his praise. Though Banion was an imposing, dominant presence, his two pupils were inspired by his forceful will. Over the months, Mineera had marveled at how much the tribals had grown in skill. Banion was a tough mentor, expecting perfection, but his pupils rose to the challenge every time. Jared and Tani both sought his earnest praise, always trying to prove themselves to him. Mineera watched Tani and Jared reenact the lesson, displaying their understanding by replicating the sequence of hand signals. Banion, too, watched carefully, ensuring that his pupils understood with perfect clarity. Finally, the display was complete. The duo had digested the lesson thoroughly. With a smile, Banion slapped the tribals on their shoulders. They looked back with a cheery gaze.

"Now for another challenge," Banion said with a cryptic look on his face. "Give me your guns," he ordered.

Without question, both tribals handed Banion their submachine guns. Grabbing one of their weapons, Banion began to manipulate the gun, pulling the slide free and removing the barrel of the weapon. Within seconds, the gun was completely disassembled and lying in pieces upon the deck of the ship.

Eyeing his feat with wild eyed amazement, both tribals looked at each other speechlessly.

With another quick flurry of movements, Banion reassembled the weapon. He was done almost within the blink of an eye, stunning them all. With a wicked grin on his face, Banion took both of their guns and disassembled them, placing the pieces in front of the tribals.

"Let's have a competition," he said with another wild grin. "When I say 'go', put your weapons back together."

Exchanging glances of surprise, Jared and Tani shrugged in agreement.

"Go!" Banion shouted.

Jared was catapulted into motion immediately. Without even thinking, he grabbed the parts and began to push them together, trying to reassemble the weapon. Tani, on the other hand, picked up the parts one at a time and arranged them in a logical order. Rubbing his chin with his stumpy fingers, the scholar tried to visualize how the parts went back together.

Jared was mumbling, still jamming the parts together. Click! The barrel of Jared's gun slid back onto the trigger housing. Tani looked over in alarm, feeling the pressure of the contest. Rattled that Jared was ahead of him, he fumbled with the pile of gun parts, trying to catch up.

Jared went on instinct alone, reattaching the stock to the weapon, and clicking it back into place. Tani was still trying to figure out how to place the barrel back on the weapon. His face turning bright red with frustration, the scholar was not amused that his friend was besting him at assembling a mechanical device, a skill in which Tani had always prided himself on his expertise.

With a final move, Jared finished building the weapon. Placing it on the deck of the ship with a triumphant grin, he withdrew his hands, signaling that he was done.

Tani sighed in frustration, looking like a defeated soldier, holding the remaining pieces of the gun in his hands.

"Good job." Banion's tone was serious. He moved over to Tani and knelt down. Taking the parts from his hands, he assembled the weapon slowly so Tani could watch. His cheeks still flushed, the young scholar bowed his head, avoiding Jared's eyes out of sheer embarrassment.

"Sometimes instinct is the only thing you have to go on. Too much consideration of an event can lead to disaster, especially in a dangerous situation. Jared relied on sheer instinct, not even thinking about the task at hand," Banion concluded.

"It was luck!" Tani snorted, still trying to patch his bruised ego.

"Bah! You're just jealous. If it weren't for Banion, you would still be sitting there!" Jared fired back with a wicked grin on his face.

"Maybe..." Tani scowled at his friend and rival. "Let's do it again."

"Fine," Jared shot back.

Banion grabbed the guns and took them apart once more. Placing the pieces in front of the tribals, he held his hand in the air, then swung it down, shouting, "Go!"

With a calm series of movements, Tani took the pile of parts and put the weapon back together immediately; it took the tribal genius mere seconds to attain his goal. Jared, on the other hand, stared at Tani dumbfounded; he had not managed to fit even two of the parts back together.

"Done!" Tani said triumphantly, slamming the sub-machine gun down on the deck of the boat.

"Now who's the lucky one?" Jared grumbled as he grabbed his pony tail in frustration and pulled it downward, noticeably agitated.

Laughing, Banion smiled at the duo. "Great job, both of you." The tribals looked back with proud smiles. They had pleased their foster father and enjoyed every second of it. They were elated by his praise, but their joy proved to be short lived. A vile creature was about to ruin their fun.

A slow clapping rose out suddenly. Advancing with a sinister sneer upon his features, Lavosi came before them, mocking the trio with his dull clapping. "Simply wonderful. The tribal trash can put their guns back together! How wonderful!"

"Why don't you crawl back into the cargo hold? Maybe you can find some of your own kind down there," Banion snorted back in derision.

"Such a quick tongue. You must be so proud of your little brats. Maybe, in time, they will be able to wipe their own asses without you!" the vile rat creature goaded Banion.

"Hey!" Jared responded with a whine. "We aren't that bad!"

Advancing steadily, Lavosi came before Jared, leaping atop a crate and staring at him with his blood red, unblinking eyes. His ashen fur bristling and standing on end, the vermin was agitated, and tried to make him self look larger, glaring at the tribal with animosity. The intimidation tactics were having an effect on the warrior. Moving a few steps back, Jared conceded a mild defeat and averted his gaze. "Anything else you wish to add?" Lavosi hissed at Jared.

Jared turned away wordlessly.

"I didn't think so!" he sneered.

"What's your point?" Tani tried to stick up for his friend.

"What's my point? I am still regretting taking such a pitiful band of misfits into the northlands. I doubt you all possess the skills to reach your goal!"

"We will make it," Banion replied in a resolute tone. "With

or without you." Moving forward, he stared into the blood red eyes of the mutant. Such emptiness and hate were in the gunfighter's eyes. Feeling fear edge into his soul, Lavosi had to avert his gaze from Banion; it was too much to bear.

"Let me remind you that you need me," the rat king muttered.

"For now," Banion shot back icily. "I suggest you leave us alone. All of us."

"Is that a threat?" Lavosi hissed as his fur rippled in anger.

"Yes," the leader of Nova 7 retorted in a confident tone. "Yes, that is a threat."

The discussion was over. Turning his back on the mutant, Banion moved toward the helm, where Mineera was still steering the ship. Tani and Jared followed without another word.

Leaving the vile Lavosi alone, they ignored his anger, hatred, and seething jealousy. And so he crept back into the darkness of the cargo hold, hiding from the violent light of the sun. After only a few minutes of exposure, his pink skin was already feeling the burning might of the sun. He retreated into the darkness; the cold was soothing and inviting as he squatted down, letting the chill, dank force of his existence flood him. Seething with hatred, he thought about killing them, one by one, a dark fantasy that had gripped his sick mind since their voyage began.

In his eyes, each of them was inferior. He knew that he could sneak up on them in their sleep and kill them one by one. But while they could be killed alone, they were strong as a team. In the course of his stay with them, the sinister vermin king had tried on many occasions to drive a wedge between them. His intent was to demoralize them enough so that he could exert influence over them. The more Lavosi abused them, the more they banded together and made him the outsider. His plan to manipulate them was failing. Normally, his bullying tactics were successful, granting him control through force and intimidation. Nova 7, on the other hand, was resisting his bullying and finding strength in one another, a reaction to which Lavosi was unaccustomed. The more they resisted, the more he hated them. This seething hatred had already become an obsession. Lavosi spent a sickening amount of time fantasizing about the demise of his new teammates.

Lavosi knew that Nova 7 needed him. Without him as their

guide, they would never reach the military installation in the northlands. Without his help, their goal of finding a nuclear weapon would be impossible. This fact was the only edge he had left; it was his only leverage to exert influence over them.

But Lavosi also knew that he needed them; he needed Nova 7. Pulling up his brown cloak, the mutant touched his chest, tracing the lump of scar tissue upon it. Thinking back to the episode in the military installation, Lavosi could still feel the pain of cold steel piercing his flesh. It was a miracle that he had survived the attack. Of the more than thirty heavily armed rat warriors sent on their original expedition, only two survived the attack within the tunnels. Lavosi was one of the lucky ones. His bodyguard had dragged the heavily wounded rat king from the tunnels as the fearsome guardian laid waste to the rest of the expedition.

Shuddering, still thinking of the horrid installation, Lavosi gritted his teeth. Bubbling hatred exploded through him once more. Thinking of Nova 7 made him jittery and angry again. For the time being, he required their skill to liberate the warhead from the tunnels and to defeat the guardian of the tunnels. Until that point, both factions in the struggle would have to rely on each other.

With a slow gesture, the rat king pulled a small radio from his pocket. Sniffing the air and listening intently, Lavosi could sense no others with him inside the cargo hold. His eyes darted back and forth in suspicion. Finally, seeing that he was truly alone, he flipped on a radio that was concealed in his pocket.

"Report," Lavosi whispered into the radio.

"We are still here my lord," the radio crackled back.

"How far back are you?" the rat king quizzed the thing on the other end of the radio.

"Only two days behind you, my lord."

"Excellent," Lavosi hissed back. "Make sure you keep far enough away so they don't spot you. It will ruin our little surprise if Nova 7 catches sight of you."

"Yes, my lord. We will keep far enough back to avoid being spotted."

"I will contact you again soon. Until then, stay silent, keep off the radio."

"Yes, my lord."

The conversation was over. With a flip of his wrist, he

turned off the device and sat within the darkness and solitude of the cargo hold. The obsession had returned. Dark thoughts mastered the creature of darkness. And so he brooded, whispering in the gloom, muttering his dark thoughts aloud. As the moments lengthened, crude thoughts emerged, thoughts filled with torment and torture, thoughts about betraying and slaughtering Nova 7.

Chapter 41
The Steel Loft

The companions were in awe. Rising hundreds of feet above the surface of the ocean, an ancient oil rig soared above them. Cricking their necks, the team members eyed the monstrous structure.

Steel girders encrusted with barnacles and other sea life clung to the wash zone, the area where the ocean waves met the steel supports.   Platforms, dozens of them, had been welded to the girders, creating a series of shelves where crude houses and shacks had been assembled. Staircases and ladders were positioned among the platforms leading to the various areas and living quarters. The Steel Loft, as it was known, was a way station on the ocean in between the lush forests of the north and the barren wastelands of the south.

A thriving culture of fisherman and traders had colonized the ancient oil rig, creating a bustling society about thirty miles off the coast of the Darken Realm. The Steel Loft was the primary oil rig in the area, the center of the oceanic society. Over a dozen other oil rigs were in close proximity, and the Steel Loft served as the central hub for them all. Well known by sea faring people, the Steel Loft was an excellent place to find merchants and exotic wares. The oil rig was a giant bazaar where a variety of rare objects could be obtained.   Traders from across the ocean would occasionally assemble to haggle and sell their wares in one of the huge markets.

Still in awe, Banion barked out an order to get the tribals' attention. "Get moving!" he exclaimed, slapping them on the back as he moved forward.

Jared averted his gaze from the strange city built atop the ancient oil rig as he jumped off the bow of the ship, holding a rope in hand. Quickly securing the boat to the mooring, the tribal grabbed another rope and tossed it to Tani. The tribal scholar secured the other side of the boat. Mineera leapt off the ship after them, landing on a wooden dock rigged to one of the support pylons.

Lavosi, wanting to protect his fragile pink skin from the hateful sun, had donned his brown robes, covering his entire body with the protective clothing. He plodded forward slowly, with only the pair of dark goggles covering his eyes peeking out from the cloak. The mutant was quite a sight, and the inhabitants had already begun to stare at him.

Examining the labyrinth of staircases and platforms, Nova 7 started to work their way upwards. Ascending quickly, the team managed to make it to the main platform, which was drenched in warm sunlight.

Mineera took a brief moment to move over toward the railing. Smiling, she felt the warmth of the day fill her. The others had already left her side to search for supplies among the vendors atop the structure.

The rusted railing was warm to the touch. Mineera supported her weight upon it, bending forward to look over the edge. Down below, bobbing up and down on the ocean waves, was a collection of boats moored to a crudely constructed dock. A number of sailing folk were going about their business, stowing supplies or offloading wares to be sold at the ocean outpost. Uninterested in their daily duties, Mineera allowed her attention to drift elsewhere.

She sighed and threw her head back, letting the warm rays of the sun heat her skin. Closing her eyes, Mineera smelled the fresh salt breeze and felt the wind push through her dark hair. As the interplay between the cool wind and warm sun tickled her face, the day felt truly peaceful. Smiling, she opened her eyes and looked at the world around her.

Mineera was startled to find a large white seagull on the railing, only a few feet away. Bleating suddenly, the bird began to shriek at her. She grinned, understanding the ocean bird was trying to beg a scrap of food from her.

"I don't have anything for you," she said in a soft tone.

The bird looked back at her, cocking its head from side to side, peering at her from the corner of its eye with its head crooked. It shrieked again, more insistently. Mineera smiled in amusement. The bird could sense her detachment, and knew all too well that a meal or scrap of food was extremely unlikely. With a nasty squeal, the gull flapped its wings at Mineera in defiance, continually bobbing its head up and down in an attempt to intimidate her. She shook her head, almost wanting to laugh at the silly creature. Finally, it flew away, leaving her alone upon the catwalk.

Amused by the day's small adventure, her attention shifted once more. It had been a while since she had seen the rest of her companions.

Spinning around, Mineera looked for the rest of Nova 7. Scanning the ramshackle buildings and corroded catwalks, she tried desperately to locate the rest of her team. Try as she may, none were in her line of sight. While scanning the buildings and patrons of the ocean refuge, Mineera caught sight of an unusual woman, crouched on a catwalk with a fishing rod. As her eyes passed over the fishing woman, a clenched sensation washed over her. The feeling was so surprising that Mineera was set on edge. As she observed the strange woman carefully, a wave of nostalgia came over her. The woman holding the fishing rod was familiar, a little too familiar. Fragmented images came into Mineera's mind. She experienced a wash of fear and panic. The woman was gripped by evil and seemed familiar. Blinking several times, the psychic was seized by an eerie vision. The fishing woman was dressed in a long sleeved shirt. The sleeves were loose, and Mineera caught sight of a black ink tattoo upon her wrist.

A tremor filled her and a scream began to rise in her throat. The tattoo was that of an ominous serpent, branded on the woman's flesh. The feeling of nostalgia and sickening emotion became clear. In a frantic attempt to warn her companions, Mineera yelled in panic, "Reaper Kai..."

Banion heard the scream, and just in time. The crash of steel rung out in the quiet air. A cargo container atop the main platform opened, its steel door slamming down with a clatter. The explosion of sound sent Banion and the tribals immediately into action, fumbling for their weapons with desperate urgency.

A mechanical cackle echoed ominously. Two full squads of

Biogtech soldiers, sixteen strong, poured from the cargo container, brandishing automatic weapons. The sunlight reflected off their pasty bodies as their mechanical red eyes searched for their target.

"There they are!" a man garbed in peasant's clothes yelled. "Kill them all!" The disguised Reaper Kai pointed at the companions.

"Get down!" Banion yelled, leveling his assault rifle at the charging squads of robotic soldiers. Taking the initiative, he knelt and steadied his gun. With hasty aim, the crosshairs of his weapon fell upon the enemy soldiers. As he pulled the trigger, a burst of gunfire broke the air with loud snaps. The high caliber rounds hit true, shattering the robotic soldier's metallic breastplate and leaving jagged wounds in their wake. Crashing to the ground, the Biogtech twitched and cackled as a plume of hydraulic fluid spurted from his prone form.

The inhabitants of the Steel Loft were screaming in panic. Two local fishermen bolted through the center of the conflict and were caught in the horrid return fire. Having taken aim, the squads of robotic soldiers opened fire, shredding both the fishermen, who were attempting to flee the scene. They crumpled to the ground as the fatal wounds took their lives.

"Move it!" Banion screamed again, retreating in an attempt to find cover. The tribals were set in motion, taking up positions on either side of him. Both brandished their submachine guns and were ready to defend themselves. Following their leader's actions, they took up defensive positions behind a steel cargo container.

A sprig of frost erupted from the raven totem around Jared's neck. Growing in intensity, the cold blast blossomed into a chilling vortex of rising energy. Noting the totem's reaction, Jared knew they had drawn the attention of some sort of psychic attack. "Get down!" he yelled.

A blazing lance of fire shot forward and slammed into the metal cargo container. Massive energy was expelled as a deafening boom erupted. The hellfire detonated against the cargo container, sending melted metal fragments raining down.

"Son of a bitch!" Banion yelled as the Biogtechs moved in for the kill. Already surrounded, Nova 7 was almost helpless. "We need to move!"

The enemy opened fire, pinning the mercenary team behind

the cargo container. One squad of Biogtechs was moving around the left side of their sanctuary, while the other squad was flanking the right. The advance would quickly leave Nova 7 between the blades of a closing pair of sharp scissors.

"Give me strength." With a whisper, Mineera closed her eyes and felt a tingle in the pit of her stomach. Yielding to the growing power beginning to course through her, she began to harness spiritual energy. Feeling the fire erupt in her soul, she opened her eyes and extended her hand. A burning beam of energy, as bright as the sun, shot forward. Like a cleaving blade of white hot flame, the potent psychic attack cut through several Biogtech soldiers, melting them and disintegrating their bodies in an energetic wash of burning heat.

Caught off guard by the attack, both the disguised Reaper Kai soon figured out that Mineera was a primary threat. "Kill the traitorous bitch! Open fire!" The Biogtechs obeyed immediately.

Undaunted, she held her hand forward. A bright sphere of white light expanded like a bubble around her. Dozens of bullets slammed into the barrier. As each bullet struck, sparks of white light flashed in the spots where they contacted the psychic barrier. Not a single bullet bit her flesh; instead, the gunfire was deflected by the spiritual barrier of mystic energy.

"Keep her pinned!" the female Reaper Kai yelled in excitement, brandishing a jagged knife. Running forward, she was eager to close in on Mineera and cut her apart with the blade.

The Biogtechs continued their assault, preventing Mineera from dropping her spiritual barrier. Knowing full well that the charging Reaper Kai could pierce her shield of energy, Mineera had to act fast. Walking backwards, she held her concentration as the constant gunfire continued to target her. Bullets ricocheted off the shield and hit the ground with a clatter as she retreated.

As she attempted to find cover, Mineera was running out of options. If she couldn't find sanctuary quickly, she would have to drop the psychic shield in order to defend herself against the charging Reaper Kai. She made her way behind a cargo container, dropping the barrier for a split second before she was assaulted.

Uttering a violent war cry, the female agent of evil slashed at Mineera. Mineera was barely able to dodge the attack; the knife tore into her robe and bit flesh. A trail of blood erupted, following the arc of the blade as it whistled through the air. Staggering back, Mineera was now purely in defensive mode, still unable to draw a weapon of her own. Fortunately, she was well trained in close combat. Dodging and weaving to avoid further damage, Mineera was elated as a blast of chill air washed across the battlefield.

Jared had closed in on their position with the blue-tinted Scar Blade in hand. As he moved in for the kill, the raven totem emitted a frosty wind, sheltering him within the psychic vacuum.

Pulling the besieged Mineera from danger, Jared burst forward, putting himself in harm's way in order to protect her from the Reaper Kai priestess.

"Foolish worm," she jeered. Thinking he would be an easy kill, the Reaper Kai was ill prepared for the encounter. Channeling demonic power, she attempted to blast the tribal warrior. The attack was short lived. At such close range, the raven totem absorbed the energy before it could even coalesce about the psychic. Stunned by the lack of effect her actions were having, she jumped backwards, desperately confused by her inability to harm the tribal warrior with her powers.

As a wall of frosty wind rolled forward, her skin prickled under the chill. The tribal ground into her, the totem around his neck vibrating violently, hungering for demonic energy. Ramming forward, Jared whirled the Scar Blade around in a wide arc. With wicked aim, the blue steel blade hit true, sheering off the arm of the disguised Reaper Kai at the elbow. Stunned by shock and pain, she was now defenseless, her knife clutched in the hand of the severed arm. The tribal warrior was unstoppable and ended her life in another flurry of attacks.

Mineera clutched her bleeding stomach and found the wound to be a mere surface cut. Nodding to convey that she was fine, Mineera gestured in the direction of the last remaining Reaper Kai priest blasting away at Banion and Tani. "Let's go get him before he kills anyone."

With a staunch look, Jared nodded back, completely devoid of fear, tense and eager for another kill. His immense courage and lack of emotion were chilling to her. It was almost as if Mineera was looking at Banion's apprentice. Bounding like a cat, Jared sprinted across the battlefield. Breaking cover, he ran full speed toward the final priest. Biogtech soldiers fired in vain trying to strike the tribal, but he was too quick. By the time he passed into view, the bullets would hit nothing but the air he left in his wake.

Seeing the wild boy charging him, the veteran priest glowered at him. With a smile, he prepared to engage the warrior at close range. Pulling a steel tomahawk from his belt, the Reaper Kai flexed his muscles and held his ground.

Launching onto the cargo container, Jared regarded the priest with disdain. Grabbing the raven totem, he felt ice form around his knuckles as it vibrated violently.

"You are no child of Ceibla Moralis! The rumors about you are false!" the agent of evil screamed in anger. "Let's hope you have more on your side than a battered wooden totem!"

Nodding at the priest, Jared held the gore-stained blade and pointed it at his opponent's chest. "You're next," he scowled, with a frightful rage filling him. The totem's eyes burned bright red, hungering for another kill, urging the youth to do its bidding. Feeling the strange pull to battle, Jared let the haunting rage settle over him. Quivering in anticipation, he moved in for the kill.

The Reaper Kai was smart and confident enough in battle to let Jared make the first move. With a burst of aggression, Jared exhaled, and a frosty trail of icy breath shot forth as the totem projected its mystical energy. Taking a step forward, Jared swung the Scar Blade with all his might. The Reaper Kai was prepared for the attack and sidestepped the slash. Jared was like an angry hornet; darting forward, he tried to sting repeatedly, making quick jabs toward the Reaper Kai's stomach. Agile and an expert in close range combat, the agent of evil remained unscathed, avoiding the attacks.

"Well placed, I will give you that." The Reaper Kai spoke with a certain level of respect.

Jared ignored his words and withdrew a step, recovering some of his strength. The quick flurry of aggressive attacks had left him winded. The Reaper Kai, sensing his momentary loss of momentum, took advantage of the situation and moved in for the kill.

Swinging his metal tomahawk, the priest attempted to crush Jared's skull. The tribal was without a hint of fear; his extensive training was almost instinctual. Yielding to the call of his reflexes, Jared avoided the attack and slashed at the priest. The Scar Blade bit flesh, mildly wounding him. Dropping the tomahawk, the Reaper Kai found himself defenseless and resigned himself to death. Kneeling before the tribal, the agent of evil was sure that his end was near.

"Well met," he hissed in a submissive tone.

Looking at him, Jared was filled with displeasure. "That's all you got?" he scoffed. The totem vibrated violently. Gathering up the tomahawk, the tribal warrior threw it in front of the Reaper Kai priest. "Pick it up!"

Stunned, the agent of evil blinked several times.

The battle was going badly for the rest of Nova 7. The two squads of Biogtechs were about to overrun his companions. Without Jared's help, Banion and Tani were in real trouble.

"Jared, what are you doing?" Mineera yelled, watching the strange situation unfold.

"Pick it up!" Jared yelled. The eyes of the raven totem burned and there was a frantic, animalistic look on the tribal's face.

Seeing another opportunity to slay his rival, the Reaper Kai grabbed the weapon and charged him again.

Jared enjoyed the fight. Dodging and parrying once more, he counterattacked, grabbing the man by the throat. Bringing his head forward, the tribal head-butted him. The Reaper Kai fell back to the ground, clutching his bleeding head, on the verge of losing consciousness. Dropping his weapon again, the demonic servant crawled around on the ground, fighting off the dizziness.

"Stop it!" Mineera yelled, trying to get the tribal to snap out of his rage.

Holding the blade over his captive, Jared sneered at him as he groveled upon the ground. "That's all you got?"

"Jared! Help them! Banion is about to be overrun!" Mineera had closed in on them. Clutching Jared's arm, she shook him desperately. Blinking several times, Jared looked as if he had just awoken from a long slumber. The crazed, feral look disappeared immediately, and he looked around in confusion. Taking a shaky step forward, he looked dizzy as his left hand gripped the wooden totem tightly. It felt so wonderful; the cold chill rushed from the ancient relic as his heart pounded in his chest.

"What?" he said in a whisper, looking at Mineera in a daze.

"Help them! They are being overrun!" She motioned toward

the metal cargo crate. Banion and Tani were nowhere to be seen. Several Biogtechs were dead upon the ground, but eleven were still trying to kill Banion and Tani.

Smashing the Reaper Kai in the head with his fist, Jared rendered him senseless, knocking him out completely. Sheathing his bloody sword, he pulled his submachine gun free and ran to help his friends. He shouldered the weapon and took aim, beginning to gun down the robotic soldiers. His aim was shockingly precise; he slew several enemy soldiers in rapid succession as the submachine gun hummed away.

Several robotic soldiers sustained hits from behind, and turned to take aim at the tribal warrior. This splitting of forces gave both Tani and Banion the edge they needed. No longer under the full attack of the remaining enemy soldiers, they were able to regroup.

Banion rounded one corner and the tribal scholar the other. Flanking the cargo crate, the mercenary duo came into firing position. As Nova 7 took aim, the Biogtech soldiers were caught in the crossfire, a triangle of death in which each team member held a point. Scattered and confused, the Biogtech soldiers tried desperately to slay the human attackers.

Banion was a fearsome gunfighter. Moving from target to target, he shot rapid bursts of gunfire, felling many foes. Tani held his ground, his submachine gun pressed against his chest. With calm sweeping motions, the tribal scholar brought his gun to bear on his enemies, and downed several Biogtechs in a wash of gunfire. Finally, Jared fired from his elevated position, spraying the remaining Biogtechs with bullets.

The intense firefight was frightening to witness. The Biogtech troops never even had a chance after the trio got into solid firing positions. Watching the kill zone with fascination, Mineera marveled at the two tribals. Instead of the battle looking as if it was waged by Banion and two kids from the wasteland, it looked as if it were being fought by Banion and his two clones. Both tribals had been honed and transformed into intense soldiers, able to hold their own in combat against overwhelming odds. Shaking her head in disbelief, she watched the trio gun down the remaining enemy soldiers.

After an additional interval of heated combat, Nova 7 had

defeated the enemy ambush.

The world can produce wonders amongst its people. Fantastic acts of survival are mixed with lethal acts of heroism. Mineera had witnessed such an act, an act in which her companions had acted as a team instead of lone individuals. It was a blessing to be alive, but the battle had also been a frightening event to behold. The carnage wrought with such speed was a testament to guile and a stubborn will to survive. No longer was Nova 7 a band of misfits. Instead, they had become a well honed team, able to survive against all odds. The fate of the world rested in their hands, and they were capable hands indeed; this was truly a team that could do the impossible.

Chapter 42
Changing the Rules

Secluded in the cargo hold, the lord of rats squatted in the gloom. The dank, musty smell of the ship filled his nostrils as the hairs upon his nose twitched in the darkness. Having fled the battle during which Nova 7 had almost been annihilated by Reaper Kai forces on the Steel Loft, he decided to avoid Banion's looks of disdain for the time being. Though they had not confronted Lavosi about leaving them to their probable deaths, Nova 7 had looked at the rat king with pure contempt. Not wanting a confrontation, he had crawled back into the cargo hold, avoiding his companions' silent wrath. There he waited in silence, feeling the boat lurching across the open waves once more.

After he felt that the boat was in motion once more and they were out of harm's way, the twisted vermin lord pulled the radio from his cloak. Flipping the switch, he turned on the device so he could communicate with his cohorts, currently following them into the northlands.

"Report!" Lavosi whispered angrily into the radio.

A moment of silence ensued. After what seemed like an eternity within the dank cargo hold, the radio sprung to life and a creature responded. "You are being tracked."

"Of course we are being tracked! Two Reaper Kai tracker teams ambushed us on the Steel Loft!" the vermin lord whispered back into the radio in an excited tone. "Are there any more following us?"

"A boat with more than one hundred Biogtech soldiers and at

least a dozen priests is heading in your direction. We caught sight of them yesterday, pushing north in the same path as you."

"That's impossible!" The albino rat spoke with disdain. "How could the Reaper Kai know of our prize? How could they know about the military installation?"

"I think their spies captured some of our own. Before we left to follow you, several of our tribe went missing. The last to disappear was Slathar, my lord."

"Slathar!" Lavosi hissed. "One of my own loyal bodyguards?"

"Yes, my lord. He knew the secrets of the north. I believe the Reaper Kai now know the route into the northlands and are moving toward the military installation."

"Damn them." The race to locate a nuclear weapon had become more complicated, and Lavosi was highly displeased. Growing quiet, the vermin king pondered the situation. The evil king of the rat tribe was a monument to manipulation. Knowing that his only hope for resolving the situation was to manipulate either Nova 7 or the Reaper Kai, he decided to manipulate both of them. "I have a plan…"

"What is your plan, my lord?" the rat on the other end of the radio responded.

"I am going to get Nova 7 to let the Reaper Kai pass them by – let the Reaper Kai reach the ruins first."

A moment of silence rode the radio waves. With uneasy hesitation, Lavosi's minion responded with a confused tone. "I don't understand."

"Our last expedition to the ruined military base was a complete disaster. The guardian in the tunnels killed over thirty members of the expedition. If it is strong enough to kill an army of our kind, armed with the best weapons of the ancients, it should be able to kill the Reaper Kai horde."

"Is that wise, my lord? What if the Reaper Kai recover the nuclear weapon?"

"Don't worry. The defenses are too strong for such scum. With security systems built by ancient geniuses, it will take another genius to unravel their puzzles and defenses. This Tani, the tribal scholar, is the only one intelligent enough to unravel the secrets. The Reaper Kai are too feeble minded to succeed."

"What of us, my lord?"

"Follow us closely. I will get Nova 7 to back off, allowing the Reaper Kai to slip by. Just be mindful and stay on our heels," Lavosi instructed, his voice soft.

"Yes, my lord. We will hold back."

"Excellent." Lavosi smiled in the darkness. "I will contact you again in a few days."

Clicking off the radio, Lavosi placed the device back inside his brown protective cloak. He grabbed his goggles, placing them over his blood red eyes. Pulling the cloak over his pink skin, he prepared to leave the cargo hold.

The vermin lord sniffed the air, then moved across the floor with silent, secret steps, plodding toward the hatch leading to the top of the ship. With careful movements, he pressed up on the hatch so that only a crack of light poured into the hold. Looking out of the opening, he saw no one.

With a deft movement, the trapdoor opened and the sly rat jumped out, rushing behind a pile of supply crates. Hiding himself, Lavosi crept forward across the deck of the ship, listening intently to the sounds around him. Voices emanating from the front of ship caught the mutant's attention. Slinking up on them, the vermin lord began to eavesdrop upon a heated conversation taking place.

"What the hell happened out there?" Banion was hot under the collar, and Jared was the source of his irritation. He was referring to Jared's vicious loss of humanity as he toyed with the Reaper Kai priest on the Steel Loft. The tribal had lost control and had almost been at the point of torturing the priest.

Jared tried to look away, reluctant to respond. With nervous movements, he fidgeted with his ponytail, looking at the deck of the boat.

"Your actions could have killed us. Do you get that?" Banion berated him once more.

"I'm sorry. I lost my nerve for a moment," Jared said with a sheepish look on his face. The tribal was fighting a battle with his own emotions, and hiding it behind a distant expression.

"Lost your nerve? You were toying with that priest. You let the enemy take another shot at you, and for what reason?"

Jared did not respond, nor did he look at Banion. The lack of response was like a jolt of fire to Banion. Rushing forward, the seasoned mercenary grabbed Jared by the shoulder and turned him so that they were looking at each other face to face.

"You've got a lot of nerve, compromising all of us out there. You will look at me and you will deal with this!" Banion shouted at the tribal warrior.

Responding to the aggression, Jared's eyes locked with Banion's. The young warrior was nervous, almost jittery, as if wracked by some inner turmoil. He shook his head, his eyes filling with shame once more. "I can't explain it, Banion. I wanted to feel something, anything. Lately, I feel so empty inside."

With a sigh, Banion conceded some ground to the tribal. Having lived a dark life full of bloodshed for so long, the deranged mercenary could relate to the youth's experience. For a brief moment, Banion even felt guilty, guilty for introducing the tribal into a world filled with despair and hollow emotion. Taking a moment to collect his thoughts, Banion tried to console the youth. "I can never apologize enough for what I did to you."

Blinking several times, Jared gazed back with a surprised look.

"It was foolish of me to bring you along on this crusade. I was lost in a haze of misery and never cared about the impact on you or Tani. Back in Rasheed, I manipulated you. As a result, you have seen horrors beyond imagination. Your soul has been tested and brought to the brink, a place where I myself have been many times in my life. But in all that time, I have seen you grow and reach untold potential. If I could do this all over again, I would still take you along, but wouldn't have manipulated you. If I could do this all over again, I would have tried to encourage you to come along."

The emotionless tribal looked at Banion with a sigh. "I feel lost, Banion. I have done unspeakable things. I keep thinking that I should feel remorse or guilt, but I feel *nothing*, absolutely *nothing*. Doesn't that frighten you? Shouldn't I feel something? Anything?"

"I don't know what you should feel. Hell, I can't even figure out how I should feel, and I have been out here, killing and living life on the edge, for more than half my life," Banion responded. "You cannot afford to get pulled down into a world like mine. You have a long life ahead of you, and the last thing I want to see is another person like me walking this earth."

With a tremor rocking him, Jared's eyes shifted back and forth. He was on edge and Banion could see the turmoil on his face. Yet the tribal held back, still unwilling to reveal his secrets.

Trying to push his strife to the surface so they could both deal with his issue, Banion pressured Jared to open up. "What's going on with you?"

Jared did not respond; instead, he looked away, still unable to deal with his raw emotions and grief. Unwilling to simply give up on the boy, Banion pushed again.

"Look at me." Banion grabbed Jared by the shoulder once more, preventing him from turning away. Jared's harsh eyes locked with Banion's, and his grim appearance began to melt away as he began to quiver. Fighting back the tears, Jared was completely vulnerable before Banion. Seeing him waver, Banion persisted.

"Are you going to tell what's going on, or do I have to throw you in the water and drag your sorry ass behind this boat until you crack?" Banion's tone was lighthearted.

Blinking several times, Jared fought back the tears and let out a sigh. "I'm not sleeping well anymore. Every time I try to sleep, I see flashes and images of battles that we have been in. The events keep playing out in my mind over and over again. The longer I think about it, the emptier I feel. The emptier I feel, the more I hunger for more conflict. In a sick sort of way, I want to fight for my life so I can wash away this empty feeling I have inside. In a sick sort of way, I want to kill so I can feel alive."

The words hit Banion hard. Having lived much of his life with the same feeling, he was struck by another wave of guilt. Shaking his head in disbelief, Banion could barely comprehend that someone else felt just like he did.

"What you are feeling is wrong, it's a terrible emotion. I know how you feel, but you cannot get sucked down into a world ruled by violence. Even though you feel that conflict makes you feel alive, it's only a fleeting emotion. The more conflict you find, the less satisfied you will become after every encounter. If you do not keep your emotions in check, the violence will consume you."

"I am so close to the edge, I'm not sure that I can keep from falling off." Jared shook his head in anger. "I don't want to feel like this. I want to rest my head and let tears run down my face, wash away my grief. I want to forget this bloody road. I wish I could feel alive again."

"You can never forget the things you have seen and done. The only thing you can do is to survive and make sure you, the real you, endures this road. If you sacrifice too much of your soul, you will be lost to the darkness. Don't walk the road I have traveled. Trust me: find faith in yourself and your friends. Do not despair, or you will not have the fortitude to survive this whole mess that we are involved in."

"Maybe we shouldn't survive this. Maybe none of us deserve to survive this." Exasperated, Jared spoke in a dull, defeated tone.

"Don't say things like that," Banion countered.

"Why? We are nothing but murderers traveling under the disguise of heroes. In the end, we are murderers, seeking to kill an entire race of people. Lavosi is right; we are nothing more than murderers." Jared's voice was saturated with loathing.

"You remember the city of Rasheed? You remember how many defenseless people died that night? What did they die for?"

Jared was silent.

Banion responded to his own question. "They died for nothing. They died because their faith wasn't the same as their enemies. We are fighting for all those people who are too scared to fight, too weak to survive. We are on a mission of mercy. If we do not succeed, thousands and thousands more will die, and they will die for nothing."

With a nod of comprehension, Jared expressed his agreement. "I'm sorry, Banion. I don't know what happened out there. I guess I wanted to feel *something,* I wanted to feel alive. I should have never toyed with that priest. He could have killed me."

"Just remember not to lose what you value most – your humanity. I lost mine a long time ago, and it is only recently that you, Tani and Mineera have taught me how to care again." Banion's tone was solemn. His words filled the tribal with hope. If a broken man like Banion could find salvation, there was hope for him, as well. "Are you all right?" Banion pressed.

Jared did not respond. His expression remained distant. He was definitely not all right.

Sighing, Banion placed his hand on Jared's shoulder. "Let me rephrase that. Are you all right for now?"

Jared nodded meekly. With a weak smile, the tribal warrior looked at Banion and spoke in a near-whisper. "Thanks."

"Sure thing, Jared." Slapping him on the shoulder, Banion smiled.

At this point, their somber conversation was interrupted. A shuffle of movement disturbed both Jared and Banion. Something was sliding across the deck of the ship only a few feet away. Startled by the sound, they approached quickly and found Lavosi, garbed in his protective robes, eavesdropping on them.

"So this is what you're about? Conflict breaks out and you hide? Jared and I were talking and you spy on us? Is this what you're about? Sneaking around like a coward?" Banion was rife with anger as he stared at Lavosi.

"It is not my place to fight your battles!" Lavosi hissed back.

"What is your place?" Jared responded.

The rat king did not reply. Irritated to be discovered, Lavosi ignored them and moved toward the back of the ship. The shunned monarch was filled with an immature hatred, an almost adolescent anger as he pulled the door open and descended into the cargo hold once more.

As the vile Lavosi crawled back into the depths of the ship, Banion returned his attention to Jared. "I will never trust that damn rat."

Nodding back in agreement, Jared scowled at the cargo hold. He paused to smile at Banion in parting before heading off toward the back of the ship.

Banion remained alone, lost in thought and deeply perturbed. Where once he would have scoffed at Jared for feeling such emotions, Banion now found himself experiencing heartfelt concern for the tribal. The empathy was unsettling to Banion, but the emotion was a healthy one. The reckless gunfighter was moving towards redemption. He felt a sick feeling wash over himself as he thought about such things. As Jared was sliding downward into despair, a miserable prison of denial and ill will, Banion was beginning to find himself again.

For the first time in a long time, Banion was happy to be alive. He was happy to be able to help someone who was beginning to lose their grip. It felt wonderful to act like a leader, a real leader who could inspire and coach his companions. Smiling in the warm sun, Banion basked in his new found role in life. It felt wonderful to feel needed and know that he was making a positive impact on someone's life. While Jared and Tani were beginning to feel like his children, Banion had begun to feel like their father. Never having had a chance to have children of his own, it was a refreshing experience for Banion.

With a heart swept by conflicting emotions, Banion sat on the deck of the ship. As the boat raced across the ocean waves, he pondered his own life and the fantastic series of events that led him to that very point.

Chapter 43
The Northlands

"I am still not convinced this is a good idea," Jared whispered in a dull tone. The tribal warrior was lying on his stomach with a pair of binoculars in hand. Hiding amongst the rubble and debris, he scanned the westward flow of the river out of the interior of the continent.

Tani, lying beside him, nodded in agreement. "I don't like this, either."

The plan they were carrying out seemed absolutely insane to both tribals. Lavosi had convinced Banion that enemy agents were close on the heels of Nova 7. The gunfighter was skeptical at first, but his skepticism rapidly turned to a resolute agreement with the vermin lord. In the weeks of traveling northwards along the coastline, they had caught sight of several ships pushing in the same direction, only a day behind them and gaining slowly but steadily. This news had put the entire team on edge, and it didn't take long for Lavosi to convince them that the ships contained Reaper Kai of the same faction that had ambushed them on the oil rig.

Knowing that the enemy was close at their heels, Banion grew determined to reach the military base first and obtain the nuclear warhead before their enemies.

Lavosi had another plan. Since his first expedition had ended in utter ruin, the vermin monarch had a different agenda. He was so convinced that the guardian of the ruins, a fearsome ancient terror, would be potent enough to drive back the entire army of Biogtechs and Reaper Kai marching on the ruins, that he suggested letting the Reaper Kai forces pass them and reach the ruins first.

Originally, the suggestion seemed preposterous. Banion had snorted and scoffed at the very idea. But the longer they traveled and the closer the enemy came, the more he wavered from his initial decision. Slowly and meticulously, Lavosi wore them all down, using half truths and trickery to make them see things his way. After weeks of such perverted subversion, Banion cracked, and the new plan was to allow the Reaper Kai to pass them and reach the ruins first.

It was this decision that had led Nova 7 to the very situation in which they found themselves now. Both tribals were convinced that the new plan would lead the team to ruin and allow the enemy to win the epic race that they had struggled so long to win, but they followed their leader's orders, nevertheless.

Crawling forward, the scholar concealed himself behind a concrete pylon, eyeing the enormous river below with a tense look.

The team was concealed upon a ridgeline on the northern cliff face, which rose hundreds of feet into the air, overlooking the enormous river emptying into the ocean only a few miles west. The sheer gray stone rock face made it impossible to climb the steep face directly. The team had taken nearly an hour to push up from the coastline and up the sharp incline.

Far from the contamination and taint of ancient nuclear fire, the northlands were seemingly untouched by the ravages of war. Enormous forests and mountain ranges spread northeast for hundreds of miles, hiding any remnants of civilization under a thousand years of growth and untamed natural influences. Tall, strong trees had overtaken the cities and towns, turning roads into groves and office buildings into entangled husks of empty rubble. As the ancient people in the cities perished, only a few ventured far enough north to carve out a home. By the time the fragments of civilization tried to organize themselves into a cohesive tribe or culture, the wild earth overpowered them with harsh conditions and fearsome predators. Only a few tribes had survived the harsh apocalyptic conditions, the destruction of civilization and the collapse of government. These remnants of civilization were a collection of primitive villages, mostly along the coastline, hacking out a living by fishing and hunting deer and elk in the nearby forests.

Secluded in the woodlands, Nova 7 had taken up a position along the ridgeline to spot incoming ships. Their intent was to spy upon the boats which had been on their heels for several weeks after the violent encounter at the Steel Loft. If the sinister mutant rat was correct, the ships were filled with Reaper Kai and Biogtechs.

An ancient suspension bridge connecting two of the cliffs had collapsed ages ago, leaving only concrete rubble and debris along the cliff face. The two tribals lay among the rubble, binoculars in hand, scanning the area below. The rest of the team members were back at the tree line, about half a mile to the west, also looking for signs of the enemy. As they anxiously waited to discover who had been following them over the past weeks, Tani and Jared talked amongst themselves.

"Don't you find it a little odd that Lavosi has so much information about the boats behind us? It almost seems like he *knows* what's stalking us," Jared mumbled, taking the binoculars from his eyes and pressing his back up against the concrete pylon.

"Yeah, it doesn't make any sense. I just wish Banion wouldn't have given in to Lavosi's demands," Tani responded, inching closer to the edge of the collapsed suspension bridge. Looking down, the tribal felt dizzy. The ground fell away sharply, and he steadied himself as he looked into the void underneath him. The shattered remnants of the bridge lay in the river below, shards of concrete and broken steel poking from the rushing water. From his vantage point at this dizzying height, the tribal was in awe of the natural world about them. The team was truly in the wilds of the Darken Realm. Few had seen the wonders of the northwestern forests of the continent. An unnatural barrier of radioactive wasteland had cut off the region from the rest of the continent, blocking all forms of ground transportation. With a sense of quiet amazement building inside him, Tani felt at peace. Thinking back to his studies of ancient history, he suddenly felt like a valiant explorer, uncovering the mysteries of a dark continent. With a small nod of contentment, he moved back and sat beside his friend from Scarskin.

"Do you actually think anyone will ever believe all of this?" Jared had let his thoughts drift away from the oppressive task at hand, gazing dreamily into space. The warm sun was touching his skin. Feeling the pleasant light upon him, Jared pulled his knees to his chest and rested his chin upon them, letting out a sigh of pleasure.

Smiling, Tani shook his head. "No, Jared, I don't think anyone will believe any part of this journey we are on."

"Do you ever wonder what would have happened if we had gone north from Scarskin? Do you ever wonder how different our lives would be if it weren't for that one decision to travel east?"

Smiling, Tani looked at Jared and pushed his glasses up his nose. "I bet Mogi is still angry. I bet he is still grumbling, complaining about 'those damn kids.'"

"I can see his face right now. His brow is probably all scrunched up and his right eye twitching. He will blink a few times, scowl, and then throw his hands up in frustration." Giggling, Jared thought about the aged weapons-master from Scarskin. The smile turned to quiet, somber emotion. He whispered, "It seems like a lifetime ago."

Tani nodded in agreement. "Not only a lifetime ago, but a different life. I was so naïve. I was so scared. Now I feel confident, resolute in thought."

"I'm still scared." Jared spoke with a distant look in his eye.

Tani looked at him suddenly, blinking several times, alarmed by the admission. "Scared? Scared of what?"

"Scared of myself," the tribal warrior confessed.

"Why would you say that? You should be proud of yourself. You have grown so much."

"Have I grown? Grown into what?" With a sinister look in his eyes, Jared rubbed the strange bird totem hanging around his neck.

Watching him touch the wicked shamanic relic, Tani was growing more concerned by the moment. "What's wrong?" he asked bluntly.

"I'm not sure. I feel as if I am turning into something horrible, something lifeless."

As the words sunk in, Tani grew frightened. Usually, his friend was lighthearted, but the conversation had taken a turn toward a dark place. Pondering the stress and haunted feeling Jared was expressing, Tani fell silent, trying to analyze the conversation. "You're not making any sense."

"I know." Jared sighed, becoming sullen and quiet.

An eerie silence fell between them. Tani felt strange and for the first time in his life, couldn't say a damn thing to his best friend. They had drifted apart somehow during the madness and adventure through the ruins of ancient civilization. The silence spoke volumes. Jared had changed. Something within him had snapped and his entire personality was beginning to slip. The once headstrong, lighthearted warrior from Scarskin was now someone very different. Feeling a strange sorrow, Tani was perplexed by the rift that had suddenly formed between them. Jared was beginning to fall apart, heading into a dark place, and his friend had a front row seat for the entire show.

Thinking of their childhood, he shook his head in despair. They had grown up together and had always been the best of friends. Now, their bond had been compromised by the selfish world, ripped by war. The damage their personalities had sustained would be a heavy cost to pay even if the crusade didn't take their lives.

Tani wished that something would happen to end the stretching, eerie silence between them.

His prayer was answered. Emerging from the dense green foliage, Banion, followed by Mineera and Lavosi, moved into view.

"The boats are pressing up the river," Banion said in a crisp, commanding tone. "Now we can finally take a look at who has been on our heels all this time."

In the valley below, two steel freighters belching coal smoke from their smoke stacks lumbered up the mighty river. As the vessels pushed out from the sea, the engines whirred and roared, fighting against the strong current of the waterway. Slowly they began to head east, up the river.

Crouching down, the members of Nova 7 began to spy upon the two boats, focusing their binoculars upon the freighters with growing awe. The decks of the ships were crammed with Biogtech soldiers. Pasty white bodies and pale complexions glistened brightly in the sun. Sizing up the forces, the companions counted at least fifty Biogtechs on the deck of each boat, and no less than a dozen Reaper Kai priests.

A hollow feeling instantly erupted in the pit of Banion's stomach, as if he had been punched in the gut. The vile Lavosi had been right; the Reaper Kai had been moving toward their objective the entire time. Seeing the immense numbers of enemy troops, Banion knew that finishing the bloody crusade would be almost impossible.

"There is no way, Banion. There must be close to a hundred Biogtechs, if not more, and a dozen Reaper Kai. We cannot hope to survive that," Tani said grimly, his brow furrowed. His stumpy hand fidgeted nervously with his wire-rim glasses as he spoke.

"We shouldn't have given up. We should have stayed ahead of them. Now there is no hope." With a scowl of anger, Jared glared directly at Lavosi, making it very clear that he blamed the rat king for their unfortunate circumstances.

"They are nothing!" Lavosi hissed through the cloaked hood which concealed his albino body from the harsh sunlight. "They will not be able to stand against the guardian."

"This thing, the ancient terror that destroyed your expedition?" Mineera quizzed Lavosi. "How about you tell us more about this ancient menace?"

With eager haste, Lavosi began to recount the tale of his failed expedition to the ancient military base.

"It rose from the shadows with such speed and wrought such terrible damage upon my brethren that I was torn and bloodied before I knew what was happening. As my blood and life-force dripped from my open wounds, all that I could hear were the screams of my tribe being slaughtered in the lonely tunnels. You want to know what lurks in those ruins? It is something born from the ancient world, a fearsome sentinel capable of destroying over thirty of my own kind in a few minutes. Other than this, I know nothing, only fragments of images and memories of pain..." Their guide was shaking his arms spastically as he spoke, gripped with emotion.

"The enemy has an army of troops capable of destroying an entire city. You actually believe this guardian can destroy them all?" Banion asked, skeptical.

Without hesitation, Lavosi responded, "Yes. The guardian will destroy them all."

"So this *thing,* whatever it is, can destroy an entire army of Biogtechs and wipe out a Reaper Kai war party? What makes you think the five of us can defeat it and recover the nuclear warhead?" With flawless logic, Tani assaulted the vermin lord with his inquisitive questions.

"The key to its defeat lies in understanding the technology

behind it.   The Reaper Kai, though strong, lack the intellect to unravel the mystery of the sentinel," Lavosi hissed, pointing in the general direction of Nova 7. "This creature was born from the ancient world. Its strength is fierce technology. You will defeat it!" he boomed, pointing directly at Tani.

Forced into the spotlight, Tani took a step back. The sinister king of the rats seemed to have supreme confidence in his abilities and knowledge of ancient technology. Lavosi's unwavering faith in his intellect was unsettling, almost intimidating. Feeling frightened by the prospect of battling against the ancient terror, Tani looked to his true companions in Nova 7. They were equally stunned, feeling that Lavosi had lost his mind.

Their confidence in the course of action they had chosen to take was fading quickly. They had been drawn into a dangerous scheme by Lavosi. Doubting his sanity and ability to gauge their mission, they felt betrayed. If it weren't for his cunning words, they would still be in the lead, at least half a day ahead of the Reaper Kai army marching in on the ruins. A sense of failure washed over them. Feeling that their world was coming apart, they were filled with dread. Not only were they behind the Reaper Kai forces, but Lavosi was also relying on the tribal scholar's ability to decipher the riddle of the ruins and stop whatever protected the silent tunnels. The success of their mission, a crusade that had been the focus of their lives for a great many months, seemed to be slipping away by the minute.

With increasing urgency and panic, Nova 7 watched in silence as the two freighters laden with enemy troops pushed up the river toward the military base and their goal of nuclear retribution. The future was highly uncertain, and the stress of the events was overwhelming. Lavosi was clearly insane and had compromised the mission. Frustration and anger filled them. The end of the crusade was not going to be simple or pleasant; it would take a miracle to enable Nova 7 to succeed in their goal.

Chapter 44
Savage Fury

The day was crisp and new. A thin layer of frost was covering the land, blanketing the forest with a white, chilling glaze of ice. Fall was in the air, and the night-time temperatures had dropped below freezing. The trees sparkled as if sprayed with glistening ice. Fog had settled over the river and shoreline, adding to the icy environment.

Rubbing his arms with his hands, Jared was chilled to the bone. Having spent most of his life in open desert, the frost-filled morning was beyond his comprehension. The humid and chilled air was almost too much to handle. The frost cut through the tribal's flesh, making him shake and shiver in the brisk air. Shifting toward the glowing embers of the dead fire, he stretched out his hands, warming them in the faltering heat.

Tani was moving around in an almost random manner. The icy morning had also taken him by surprise. He was darting back and forth in an attempt to warm himself, circling the camp and trying to get his chilled blood pumping through his veins. From time to time, he would remove his wire-rimmed glasses and wipe away the frost which had accumulated upon them. All in all, the tribal scholar was miserable.

Banion and Mineera were assembling their belongings, getting ready for the rest of their journey. According to their sinister guide, a small outpost was located only a half day's journey away, at most. With the onset of fall and their northern position on the continent, the environment was already becoming hostile. Their intent was to grab fresh supplies and winter clothing, then push on toward the abandoned military base.

The group's collective mood was sullen. Each of them was fighting off the anxiety pressing against their frayed sensibilities. The dread forming in each of them centered on the race which they were now losing. Enemy forces had taken the lead and would reach the ruined military base first. The only hope they had left was a cryptic description of a fearsome guardian which had destroyed Lavosi's first expedition. The sinister rat king was convinced that whatever lay within the abandoned tunnels would be more than enough to slaughter the Reaper Kai agents; the other team members were not so certain.

Propelled by an anxious need to get moving, Jared stood up and grabbed his gear. "Are we ready to go? The day isn't getting any younger."

Banion pulled his hat back to keep the chill off the back of his neck. "I agree. We need to get moving. I don't want to fall too far behind."

Tani and Mineera nodded in agreement. Grabbing their own gear, they moved toward Banion. Lavosi, standing upon a rock near the river's edge, was ready as well. He was looking eastward, up the river, as if planning their travel route. The vermin lord was silent; he had become more reclusive since the team left the ship. In contrast to his obnoxious and sometimes downright nasty manner on the boat ride north, Lavosi was now keeping to himself, trying not to agitate the rest of the team. His uncharacteristic actions had begun to worry the rest of them. The vermin lord had become very serious and calculating. It was apparent that Nova 7's failure was not an option; the manipulative monarch would see this mission through to the end.

"Follow me," Lavosi hissed, launching off the rock and moving along the shoreline of the massive river with quick steps.

The sky was overcast, hiding the sun behind a thick wash of gray clouds. In this dingy environment, the pale albino rat had removed his hood. With pink ears rising from his head and blood red eyes staring ahead, he was a ghostly sight as he moved through the fog. The team followed their leader a mere twenty minutes into the forest before the rat became startled.

Stopping suddenly, Lavosi brandished a submachine gun,

holding it ready as he sniffed the air with his long pink snout. His ears twitched as if probing the forest around them. Bringing the slide back, the mutant rat chambered a bullet in his gun.

Banion was set on edge by his actions. In their entire travels, Lavosi had never drawn a weapon. Not trusting the rat, Banion readied his own weapon. If the vermin king was about to betray them, he would be ready. Seeing the tension on his face, the tribals fumbled for their own weapons. Mineera pulled her blue robes about her, probing the rat lord with her potent psychic abilities. She could tell he was on edge, but couldn't discern why.

The pale white hair began to stand on end upon Lavosi's head. Dropping down, he continued to sniff the air, his features pinched and taut.

"What is it?" Banion quizzed him, concerned, as he scanned the tree line for possible hostile attack.

"Cannibals." Lavosi spoke in an ominous tone. "I can smell them. About half a mile from us, somewhere in the trees. Their scent is faint but I can smell them nonetheless."

"Cannibals?" Jared quizzed. "What the hell is a cannibal?"

"Someone that eats their own kind," Tani shot back grimly.

"Eats their own kind?" Jared was aghast. "People that eat humans?"

"Yes," Mineera replied, whirling around. Her light blue eyes scanned the fog behind them as if to catch sight of a shambling menace in the haze. Nothing, absolutely nothing could be seen in the dense mist.

"How do you know?" Banion was skeptical. "How can you tell cannibals are out there?"

"On our first journey, we came across a fur trapper camp. The trappers had been killed and stripped of their flesh. All that remained was their dismembered bodies. They were covered with crows, squawking and feeding on their remains. Forms lingered in the foliage. As we passed, we could see them, dressed in animal hides, covered in blood, many still gnawing on the bones of the slain trappers. They had a noxious smell, the same scent I smell now. The odor of death is unmistakable."

"Did they attack you?" Tani asked in concern.

"No, no," Lavosi said as he pushed forward, wanting to make haste. "We were large in number, thirty strong; they didn't dare attack us."

"We are but five." Mineera's voice was haunting. A shiver shot down her spine. Her psychic powers had picked up on the emotions of the primitive, sub-human creatures dwelling within the trees. A feral, primal rage flooded her mind. Blinking several times, she found herself submerged in an image of animalistic humans feeding upon warm flesh. She shook off the hideous image determinedly, uttering a silent prayer.

Picking up the pace, they moved quickly toward the eastern headwaters of the river. The fog became a cloying mass of shadow. In the thick haze, the companions could only see a few feet in front of them, and had to rely solely on Lavosi's keen senses. Following closely, they heard nothing beyond the muffled roar of the river echoing in the white darkness.

The humid, chilling fog froze them to the bone. Shivering, each of them could see the frosty breath exit their mouths. They continued to plod silently after their wicked guide, trying to ignore the frosty conditions which seemed to add an extra weight to their footsteps.

Sniffing the air, Lavosi readied his weapon once more. Something was close, very close. A rustle of movement reached their ears. Anticipating an ambush of some sort, they prepared themselves for the worst. In the thick fog ahead, a hunched form rested upon the beach. With caution, they all moved forward, weapons at the ready. The thick mist forced them to get dangerously close to the unknown entity in order to get a good look at it.

A wasted form, barely human, was gradually revealed. Brown eyes looked out from a mask of dirt and filth. Covered head to toe in mud, the primitive human sniffed the air, catching their scent. Long hair covered its body and foul smelling rags made of animal hides hung about its waist, forming a crude loincloth. Scanning them with its frantic, hungry eyes, the feral human grunted.

Nova 7 stood their ground. The two factions faced off. With an angry stance, the primitive human rose from the ground and flailed its arms. With a howl of rage, it jumped up and down, smacking the palms of its hands upon the ground. Slapping the earth, it rocked back and forth, grunting in a savage tone. Each time it grunted, it moved forward a bit, trying to intimidate the group.

With a host of guns at the ready, the primitive human was sorely outmatched. But with each grunt, the team was wasting precious time. The noise the savage human emitted was echoing through the fog, traveling miles in every direction.

"We need to move!" Lavosi hissed, turning around and sniffing the air. The others were startled and unsettled. A devious, frantic look covered Lavosi's face. Without warning, he opened fire, with immediate and drastic results. Several bullets struck the savage human, which fell upon the rocky shoreline, its life-force fading as blood spewed from the open wounds while the gunshots echoed in the mist.

"What are you doing?" Mineera asked sharply. The attack was so aggressive and heartless; it filled her with a sickening dread.

"Saving our lives," Lavosi hissed.

The forest seemed to come alive. Grunts, strong and numerous, echoed from the trees. A host of savage humans had been summoned by the call of their dying brother. In the thick fog, dark shapes were fleetingly revealed as they charged toward their position.

"We have no advantage. Our guns are useless in this dismal fog. Run for it!" Banion urged and began to charge down the beach. As he disappeared ahead of them, the rest of the team followed the sound of his feet striking the rocks on the stone-littered beach.

A howl erupted to the right of them. A shadowy form screeched and lumbered out of the fog. With arms outstretched, the savage tried to grapple them. Jared sidestepped the beast with a confident move, swinging his left hand. His fist smashed solidly into its jaw. Howling in pain, it crashed to ground.

Tani and Mineera were both a step behind the tribal warrior. Seeing the barbaric human collapse, both jumped over the fallen wild man. The savage had fallen, but many more forms roared from the fog, trying to cut a swath through the fleeing team. Dodging the outstretched hands, they ran for their lives, aware of the consequences of even one wrong step.

A burst of gunfire rolled from the fog. Lavosi had fallen back, his short legs causing him to lag behind. Unable to see where Banion had gone and with gunfire ringing out in the white fog, Tani,

Jared, and Mineera stopped and looked around in confusion. Shouldering his submachine gun, Jared drew the mighty Scar Blade. Forming a semi-circle, the three companions scanned the white darkness for incoming foes.

The clap of gunfire was drawing the host of primitive warriors. Ignoring the trio, the horde of dirty humans rushed toward the sound, attracted to the noise like moths to the flame. Standing in the white darkness, the trio heard the strains of combat in the distance.

Grunts and screams rang out, sickening groans of death mingling with the roar of bullets. Breathing heavily, the trio was paralyzed by the resounding carnage. A gagging sound erupted and turned to a whimper. A snap broke the air as another bullet sounded. The grisly cries of war were abruptly cut short. Another burst of gunfire tore the air. A piercing scream, one laden with suffering and pain, ripped through them. And then, as quickly as it had started, the sound of combat ended. An eerie silence washed over the river valley.

The team stood silently, hearing their own hearts pounding and the sound of the nearby river churning violently. Their ears straining, each of them probed the fog for aural clues.

Their senses on overload, the three companions were stunned by a form emerging from the fog. Spinning toward the movement in fright, Tani almost shot the shadowy form. A second passed, and Banion emerged from the fog. His taut finger stroking the trigger on his assault rifle, he looked in concern at the rest of his team. They all appeared unharmed, which calmed Banion considerably. As the team assembled, it was quickly evident that the sinister rat king was nowhere to be seen. Each of them knew Lavosi was battling for his life in the fog.

Another shriek erupted from the thick mist, quickly turning into a frantic panting, like a dog wheezing after chasing a rabbit. The wheezing turned into a gagging sound which dissipated gradually. As the sounds of agony fell away to a trickle, an eerie silence filled the air once more. The battle had ended and none of them knew the victor. As their eyes strained downriver, a small form emerged.

To say that Banion, Jared, Tani, and Mineera were stunned to witness the twisted vermin king emerge from the fog would be an immense understatement.

His pasty white fur spattered in red blood, Lavosi moved toward them with a manic look. He appeared almost elated by the slaughter he had wrought with such speed. Eyeing him in revulsion, the original members of Nova 7 stared in horror at his mouth. The vermin king was breathing heavily and blood, sickly red, was dripping from his razor sharp teeth. Lavosi had been so caught up in the frenzy of combat that he had bitten into one of his enemies like a rabid dog. Even Banion, who had a lifetime of wicked memories, was shocked by his appearance. Still reeling from the memory of the horrid sounds of battle, the team members were taken aback by their newest member.

Lavosi was not to be underestimated. He was a vile, cruel, manipulative persona, brutal in both tactic and thought. At that moment, their reservations became almost overwhelming. Their guide was nothing more than a homicidal maniac, bent on bloodshed and suffering. His actions cast serious doubt on the chances of completing their crusade. Banion and the others knew that they had to keep one eye on the Reaper Kai and their other eye on Lavosi.

Breathing in quick gasps, they viewed their guide in shock. Wiping the blood from his mouth, Lavosi gazed back with a heartless, predatory expression. His voice was a near-whisper. "The outpost is near. Let's go."

The sinister rat moved into the fog. The rest of Nova 7 remained in their places for a brief second. Glancing at each other in shock, each of them knew what their companions were thinking; the end of the road would not be easy.

Chapter 45
Great Hearth Lodge

White flakes began to fall from the gray sky. At first, only a few lonely flakes floated downwards from the heavens. Gradually, more and more joined them. The tribals were in absolute awe. Never before had they seen snow, or even dreamt of such a thing. The harsh desert environment in which they had grown up barely allowed rain to fall from the sky, let alone icy snow.

Extending his hand, Tani let the flakes hit his bare skin. He watched in fascination as each flake melted instantly on his hand. A broad smile graced his lips as he watched intently. Throwing his head back, he allowed the light snow to land on his wire-rim glasses. The delicate flakes hit, melting immediately and leaving dots of moisture on his spectacles. Looking out through his spotted spectacles, the tribal grinned again.

Jared was also elated. He watched the world around them turn into a raging blizzard within a matter of moments. The dark sky had ushered in layers of gray clouds, billowing like smoke from a rising fire. Each clutch of clouds seemed to get closer to the ground as the violent snowstorm pushed closer to the travelers. In the distance, just to the west of them, the trees disappeared into the billowing wash of falling flakes. A white haze surrounded the forest, engulfing it in an icy fury, which first resembled fog, smothering the green trees. As the world seemed to be devoured by the storm, fierce winds erupted from the surging clouds.

The scarce snowflakes were no longer sparse. A torrent of heavy snow with large flakes was pushed ahead of the mighty storm cell by the raging winds. The flakes were being pushed vertically by the gale-force winds.

The blizzard soon turned the initial fascination into worry. Within minutes, the raging storm had engulfed the forest in frightening white-out conditions. The temperature had dropped significantly, making the team feel as if they had all stepped into a freezer. Barely able to see twenty feet in front of them, Nova 7 slowed their progress considerably. Banion halted his advance. Turning around to view his team, he quickly gathered them around him. In the fierce winds, the rancher had to yell to ensure his companions could hear him. "Stay close! Don't let the person in front of you out of your sight!"

"We need to find shelter!" Jared was shivering already, also shouting in the fierce wind. The tribal was unprepared for such an environment, and his clothes were insufficient to stop the brutal chill. In his thin cloth outfit, Jared was wishing that he had something more substantial.

"I'm freezing already and the storm just started!" Tani complained loudly, also shivering. The scholar's spiky hair was covered in a thin layer of snow. With pale flakes clumping to the spikes in his hair, the tribal scholar looked somewhat like a white porcupine.

"Lavosi!" Mineera called out, looking at the vile mutant. "How much further to the outpost?"

"Another hour at most, probably less." His voice was a whisper, nearly drowned out by the howling wind.

"Let's go. I don't want to spend any more time out here than we need to," Banion ordered, allowing Lavosi to move to the front of the formation. Pressing on down the worn trail, the team walked single file, leaving only a few feet between every two members.

The world around them became utterly hostile. As a stream of humid air drove the storm, heavy snow had already begun to stick to the ground. Within the span of ten minutes, the earth around them had turned completely pale in color. It was as if the land had been covered with a white, pristine blanket. The trees disappeared, stripped of their color and form as the heavy snow clung to every surface; only shallow pockets of color remained visible.

Lavosi led the way slowly, his short legs pushing forward. He struggled in the violent wind, falling to his knees on several occasions. As the fury of the storm increased, the trail had become obscure at best. Using the river as his only real landmark, Lavosi led the team eastward, knowing full well that the wilderness outpost was very close to the river. Getting too far from the river's edge could lead to disaster. With the violent wind and billowing snow, getting lost in the forest was a real threat. One false move could result in all of them freezing slowly to death in the harsh environment.

Jared was flexing his hands in the chill air. The wet snow had already soaked the warrior, stripping the last remaining heat from his body. His fingers were completely numb. Frightened by the prospect of losing his fingers to the cold, he was trying to force blood into them by flexing his fists. With each squeeze, his fingers burned intensely. The blood rushing into his pasty white fingers felt like fire. It was an odd sensation, the numbing cold being driven away by a fresh warm wash of blood flowing into his fingers. Even though he knew the rush of blood was going to keep his hands from freezing, it felt like hundreds of needles piercing his fingers as the heat fought back the frostbite.

Tani, on the other hand, had buried his hands in his armpits. Having studied ancient medical textbooks, the tribal knew that under frigid conditions, the body would force blood toward its center. Embracing the wisdom of the ancients, he had placed his freezing hands in his armpits to warm them. His plan seemed to help; the scholar had avoided the first stages of frostbite, unlike his companion from Scarskin. Motioning to Jared, Tani demonstrated the warming tactic. The warrior nodded back and followed his lead.

Seeing the tribals struggling, Mineera fell back a few paces. Extending her arms, she urged them toward her. Seeing her gracious move, the tribals eagerly rushed forward and flanked her. Like a mother bird, Mineera draped her blues robes around them and huddled them tight. The tribals accepted the charity gratefully, each of them nodding in appreciation. Mineera simply looked back with a smile. Her blue robes offered considerable shelter from the storm, effectively blocking it and allowing them to avoid a further coating of wet, heavy flakes. As the biting wind abated significantly under her protection, the tribals felt that they could make it quite a bit further.

Windy gales broke over the forest, nearly knocking all of

them to the ground. They fought to remain standing as the storm increased its violent fury. More snow, even thicker than before, poured from the gray clouds. The wind changed direction rapidly, blowing snow around them like a tornado. In the chaotic wind, it was hard to focus. On several occasions, the team became confused and walked in circles, finding their own tracks in the newly fallen snow.

After what seemed like hours in the blizzard, a vision of hope appeared before them. Barely able to see ten feet in front of them, the team was elated to catch sight of an enormous barricade looming from the white-washed world.

Giant trees had been honed to needle-sharp points, which were jutting from the ground at a forty five degree angle. Beyond the spiky barrier was a thick wall constructed of chopped trees, standing just under three stories in height. The entire perimeter was made of hundreds and hundreds of sharp stakes pointing outward like a pincushion. Whoever had constructed the strange barrier had made it clear that it was intended to keep aggressors from scaling the wall. Anything attempting to climb the wall would be impaled or shredded to bloody ribbons by the sharp wooden stakes.

Lavosi nodded in acknowledgement; they had reached their goal. As they approached the southern edge of the spiny barricade, an enormous wooden door was revealed in the wall of rising timbers. A steel bell was mounted upon a wooden stand to the right of the gate. A mallet with a polished stone hung next to the bell. Grabbing the mallet, the vermin king swung it at the bell. Resonating in a dull tone, the bell vibrated and announced the arrival of Nova 7 to those who resided within.

A man with light red skin, dressed in animal hides, peered over the edge of the wall above the gate. Seeing that evolved, civilized beings were down below, rather than savage, humanoid creatures, he waved in acknowledgment. After a few seconds, the door was opened, allowing the team to move inside the wilderness outpost.

A feeling of accomplishment washed over each of them as they entered the fort. The red skinned man mumbled in some unknown language and pointed toward an enormous lodge at the center of the walled fortress. Seeing the monstrous building in the haze of snow filled Nova 7 with hope. Smoke was billowing from several smokestacks on top of the mighty building. Knowing that a nice warm fire was inside the building, they rushed toward the refuge, eager to escape the hostile storm.

The path leading to the lodge was constructed of flat stones, painstakingly placed to fit perfectly with one another. It was a beautiful display of craftsmanship that provided a mere hint of the impressive lodge that lay beyond. Moving tersely, the team pushed open the door leading into the enormous structure. As the wind stung their eyes, the warmth of the lodge seemed to radiate outward in a blast of peaceful, welcoming heat. Rushing inside, they forced the door shut, blocking the chill air from their presence.

With a sigh, the half-frozen team of bedraggled mercenaries shook off heaps of snow onto the floor of the lodge. Heat, forged from not one but four blazing fires, forced away the icy chill of the storm. The interior of the lodge was dark, and it took a brief moment for their eyes to adjust to the dim light. After becoming accustomed to the shadows, the team members gasped in amazement.

The lodge was constructed entirely of wooden timbers, lovingly arranged in a hexagonal pattern. Wooden furniture, including beds and dining tables, was aligned against the perimeter walls. Cushions and warm blankets woven by local craftsmen were adorning the finely crafted furniture. Soft reds mingled with the dark wooden beams, giving the inside of the lodge an earthy, warm feeling. It was the kind of place in which one could curl up in a blanket and sleep the night away, never wanting to leave.

The center of the room was dominated by an enormous circular table made of dark wood. Around the table, an assortment of locals converged; native tribal people with light red skin intermixed with white skinned fur trappers and merchants from distant lands. Four live fire pits surrounded by rock hearths flanked the table, their orange flames leaping into the air. Above each fire pit was a steel device shaped like a bowl, used to trap the rising hot air. The rising heat and fire were channeled into enormous steel ductwork that ran into the ceiling. Holes had been cut in the ductwork at regular intervals, and heating vents radiated the warm flames of each hearth along the walls of the lodge. It was a wondrous work of engineering that kept the entire lodge warm even in the fiercest winter storms.

Already awestruck by the wondrous lodge, the companions had yet to observe the final and most amazing aspect of the building. At the center of the room, above the middle of the round table, was a finely crafted skylight. Six pieces of clear glass rose and met at an apex in the very center of ceiling. The rising heat of the lodge collected in the pocket of glass, heating its surface. Even though the blizzard outside was heavy, covering everything in a thick white coat, the heated glass in the sky light was melting the snow and even evaporating the leftover moisture. The result was a perfectly clear portal into the world above. The gray storm clouds could be seen clearly through the wondrous skylight.

The surreal lodge, tucked away in the middle of a vast and unexplored wilderness, was an amazing sight to behold. With a smile, Tani looked at his companions. He rubbed the last remnants of snow from his head and wandered over to one of the mighty hearths. Kneeling down, he warmed his hands in the glow of the orange flames. As he rubbed his left hand with the stumps of his fingers, the tribal felt truly alive. His open smile was an inviting sight. The rest of Nova 7, Lavosi included, approached the same hearth and warmed themselves.

Jared's fingers were red at the tips, after frostbite had taken a brief hold over his hands. Rubbing them in the warmth resulted in another explosion of pinpricks. The jabbing pain was fierce, but he knew it to be a good sign. His fingers responded immediately to the flames and began to spring back to life.

Mineera pulled back her hood. With a grin on her face, she removed her leather boots and warmed her feet by the fire. Banion threw his black cowboy hat down and grabbed a chair. Propping up his feet in their combat boots, he warmed himself by the fire. Even Lavosi was content, pulling the protective brown cloak away from his body and letting the heat touch his frail, ashen form.

The patrons of the wilderness fort were peaceful people of the wilds. Most of them were inebriated, having partaken in fine beers, and were lazing the afternoon away, spinning tales of the wild land about them. The lodge was a place of serenity, a haven from the dark world. With danger a distant thought, the team soaked in the harmony of the bastion of goodwill. In a world ripped by war, it was a true wonder that such a place existed, tucked away within the trees, at the very edge of civilization.

For the moment, they were at peace, safe from the wild storm blanketing the earth beyond the wooden timbers. It was a small blessing, feeling the heat from the fire. With exhaustion guiding their weary eyes, they sat in silence, watching the orange flames eat the wooden logs burning brightly within the stone hearth. As drowsy waves washed over them, the team fell into a much needed slumber around the warmth of the peaceful hearth.

Chapter 46
Thirteen Moons

A cackle erupted. Jared sat straight up, fumbling for the mighty Scar Blade. His hand instinctively clutched the hilt of the mighty sword. Without thinking, the tribal warrior had pulled the blade half way from the scabbard before he regained his bearings. Blinking several times, he caught sight of the source of the cackling. A gangly man with a full beard was slapping his hand on the grand wooden table in the center of the great lodge. An empty beer mug was tipped over in front of the raving lunatic.

Relieved, Jared sighed and looked at his right hand. The blue blade was glimmering in the dim light of the lodge. With a sense of sadness, he sat motionless, watching his quivering hand. It was trembling severely, and not out of fear. It was as if he was anticipating something, something horrible. The youth was, in fact, beginning to lose his grip on reality. He was becoming consumed with a dark obsession. In quiet times, Jared would envision battles, horrible battles where he would be spinning and cutting his way through a host of enemies. These thoughts were his primary focus, and were beginning to take their toll on his mind. Still unable to feel emotions of sorrow or guilt, he knew that if he did not recover his sanity soon, he would be utterly lost, consumed by Banion's dark crusade.

The bloodlust was beginning to fade as he looked around the room. His quivering hand pushed the ancient sword back into its scabbard. For the time being, the blade would be quiet, at peace.

Sighing, Jared shook his head in frustration. His thoughts

began to wander and he soon became irritated. Unconsciously, the tribal had been toying with the strange raven totem around his neck with his left hand. As of late, this mindless habit had become a strange tick. Pulling his hand away, he tried to exert self-control, telling himself that he needed to stop caressing the wooden totem. But no matter how hard he fought its pull, he would find himself several times a day gripping it tightly, feeling its worn surface against his skin.

The totem was strange. Jared could have sworn that it bore some sort of presence. It was elusive, only making itself known from time to time. When the tribal felt it to be near, he could almost feel emotion of some kind sting his mind. In times of strife, the wooden relic not only protected the youth from evil energy, but also seemed to enhance his senses. Jared was convinced that in the midst of combat against Reaper Kai, he could anticipate his enemy's moves before they even acted. This heightened combat prowess was addictive. It was this unsettled feeling of simmering rage and lightning-fast reflexes that made him hunger, crave the bloody strife of war.

Though the totem was an object of power, it was born from the spirit world nonetheless. The willful energy was foreign to the living, and merging one's soul with the netherworld was a dangerous prospect. Jared was now walking the line between the living and the dead, fueled by potent ancient magic.

From time to time, the youth had wondered if the totem was evil. He had discarded this idea as pure nonsense. The totem itself wasn't evil; instead, it hungered for dark power. In order to satiate its desire for dark energy, the tribal youth had to walk a dangerous road, encountering evil with which to feed the totem. With each encounter with the Reaper Kai, Jared seemed to grow stronger, frighteningly more powerful. The totem seemed to be transferring some its captured energy directly into the young tribal. A relationship had formed between Jared and the living totem. They were now both symbiotically joined, mutually feeding each other's desires.

Lost in thought, Jared gripped the raven totem tightly. It emitted a mild chill as it always did. Only during times of strife did the chill rise into a swirling maelstrom of spiritual energy. As he rubbed it with his thumb, it responded to his caress. A presence moved closer.   Closing his eyes, Jared became intoxicated once more.

Rocking back and forth, he was lost in a dark trance.

"Hey," Tani said in a drowsy tone.

Jared did not respond and kept his eyes closed, still rocking back forth with both hands now clutching the totem.

"Jared?" Tani spoke again, this time grabbing his friend's shoulder and shaking him fiercely.

Coming out of the strange trance, Jared looked at Tani with a confused gaze. The scholar's brow wrinkled in concern. "You all right?"

Stammering, he shook his head, driving away the fatigue. Dropping his hands from the totem, he tried to collect his thoughts. "Yeah, I'm just a little... sleepy."

"Sleepy?" Not buying his best friend's lie, the scholar's tone held a hint of sarcasm. "You ready or not?"

"Ready?" Jared was mystified. "Ready for what?"

"Are you serious?" Tani was taken aback by his friend's indifference. "You don't know what tonight is?"

Shaking his head in confusion, Jared simply stared, dumbfounded.

"It's a full moon," the tribal scholar answered his own question.

Hearing the answer, Jared sprang to life. Staring around frantically, the youth could tell that night had fallen since they had arrived in the wilderness fort. Their afternoon nap had lasted into the night. "Damn it! Why didn't you say so?"

"I thought you knew."

"I just lost track of time, is all. We have been out here a long time."

"Yeah, you could say that. Tonight is the thirteenth full moon. Only seven more to go and we can return home." Tani spoke with a triumphant tone of hope.

"Well, damn, let's see if we can get a hold of some beer."

"I thought you would never suggest that. Hell, the way you were acting, I thought you were drunk already."

Smiling, Jared felt better. The drowsy, muddled thoughts had lifted. Seeing the drunken fur trappers and local inhabitants around the enormous circular wooden table, he knew that a supply of beer was close. Excitement gripped the youths from Scarskin. Jumping to their feet, they set out to explore the lodge. It didn't take long for the tribals to catch sight of a cask of beer which was already tapped. Clean beer mugs were set near the keg. As they looked around for some sort of barkeep, a local mumbled something unintelligible and motioned for them to take a mug.

That was all the invitation Tani and Jared needed. Filling their mugs, they moved over to the round table and eyed the motley assortment of patrons with uneasy glances. Seeing the two youths seat themselves drew the attention of everyone at the table. With quick glances, their eyes examined the heavily armed tribals. The two were literally covered in weapons, which made some of the assembled patrons somewhat uncomfortable. The wilderness fort was well off the beaten path, and war never a topic of conversation. The real world was too far away to be a real concern in the isolated northlands. Seldom did travelers from the southern reaches pass this way, and when they did, none had the level of armament which the tribals were toting.

Sipping their beer, the companions eyed the other patrons around the table with silent terseness. A grizzled fur trapper sized them up and spoke with a backwoods twang in his speech. "Where did you two little skunks come from?"

Tani looked at Jared, who returned his gaze. Taking the lead, Jared responded softly and carefully, knowing that everyone in the lodge was listening. "From the south. We're on a tour of sorts."

"Tour? Why the hell would anyone take a trip up here? Nothing but bears, trees, and feral humans."

A red-skinned tribal laughed dryly, watching Jared very closely. The wizened man could tell that the two youths were not telling the entire story. "Not many tour the north with automatic weapons."

"They are for protection," Tani replied in a sheepish tone.

"Protection from what?" another patron pressed, staring down the young scholar.

They both fell silent. Neither was about to tell the other patrons their true intent in the northlands. Avoiding the glances of everyone in the lodge, the tribals from Scarskin buried their heads, staring into their beer in the lengthening silence.

"You're going to the ruins, aren't you?" a red haired man

who hadn't had a decent haircut in years asked confidently, his gaze boring into them. "The only ones with guns like yours up in these parts go to the ruins."

"I don't know what you're talking about. We're just passing through," Jared responded in a matter-of-fact tone.

"That so? Where ya headed, then?"

Silence followed. Jared was too naïve about the northlands to even make up a convincing lie. Maintaining his silence, the tribal warrior drank a heavy swig of the beer.

"Take my advice, stay out of there. No one that goes in ever comes back out. That place is cursed," a trapper proclaimed with dark certainty.

"Like my friend said, we are just passing through," Tani corroborated his friend's lie.

"Well, that's great, then. I would hate to see you both get killed by that damn ghost."

"Ghost?" The topic had drawn both tribals' interest, and Tani responded immediately. "There are no such things as ghosts."

The local tribal man chuckled, folding his arms across his chest as he eyed the foolish scholar. "I thought there were no such things as ghosts, too. I thought the local legends were just stories told to scare little children. I was sorely mistaken." Picking up a mug of beer, the red-skinned man took a swig and his eyes grew soft. As he stared into one of the nearby hearths, the orange flames reflected in his eyes. With a sigh, he looked back at the tribals, his gaze solemn.

"I was just a boy. My father was a seasoned trapper, and it was spring. The bears had just awoken from their winter slumber, so we moved from the valley to avoid their frightful hunger. My father had taught me to stay away from any bear in the springtime. Their hunger from the winter is mighty and they will kill any man to feed their hunger. Keeping away from the lowlands, we pushed toward the shadow of the mountains to the north. We spent many days in the wilds, looking for deer and elk to hunt. With herds scarce, we pushed ever northwards, along the foothills. On a mountain pass leading to a sheltered valley, we came upon a mighty herd of elk. We tried to sneak close enough to fell an elk for food, but in the confined mountain pass, the herd was skittish and fled north through the pass into the valley beyond. I wanted to give chase, but my father held me back, saying that the valley beyond was cursed and filled with death."

With a dreamy look, the local tribesman remembered his distant past, so long ago. "I was young and proud, in my fourteenth year of life. I scorned my father's fear and scoffed the legends of old. With an arrogant heart, I left my father's side and gave chase, following the herd that had passed into the secluded valley. For hours I gave chase and finally got close enough to the herd. It was nighttime when I tried to make my kill. Crawling on my belly, I made my way through the forest glade toward the mighty herd. When I was close enough to shoot one with my bow, a fearsome sound erupted in the dark forest. It was a scream of some beast, or creature. The piercing wail stunned the herd and they fled from the glade, almost trampling me. I took shelter and waited for the final elk to flee." The man fell silent. With a tense look, he halted his story and shook his head, fighting back some inner fear.

"What did you see?" Jared gasped, absorbed in the tale.

"I don't know... I still don't know. A light, red like the sunset, a lonely light was moving in the trees. It was hovering off the ground, veiled by shadow. It was as if some enormous form was moving through the trees and the glow was held aloft. I was filled with fear and froze in terror. A grinding sound arose from the beast, as if corroded steel was being pounded by mighty hammers. The light came closer and closer to me. With my heart in my throat, I ran through the dark forest and did not stop until the sun rose the next day. I returned to my father, who had made camp in the mountain pass. He never said a word to me, but I knew he thought me foolish. I have never been back to that cursed valley in all my years."

Nodding in agreement, all of the local patrons of the mighty lodge acknowledged the story as if it held great wisdom. There was a guarded look in their eyes. Even if the stories were false, all of the locals believed them to be true.

The tribals were silent, pondering the chilling story. That juncture of silence was when it happened: a mystical event of profound significance.

The skylight in the center of the lodge was lit with a bright flash of white light. The clouds forming the blizzard had swirled and broken apart just long enough for the holy glow of the full moon to show brightly. With the earth blanketed in snow, the light was like a beacon, lighting the world, reflecting off the newly fallen snow. The soft white light poured in the glass skylight, bathing everyone around the wooden table in an eerie flare of ashen glow. The strange event turned the entire room into a silent, awe-inspired group of dreamers. Seeing the clouds part and the full moon blast the darkness in fiery white light was a unique occurrence. Everyone in the lodge was mesmerized.

The youths from Scarskin felt a chill run down their spines. The event seemed like a powerful omen. Smiling, watching the perfectly round form of the full moon through the skylight, the tribals were filled with a twinge of hope. Feeling the white light upon their faces, they threw their heads back and stared at the holy full moon. Many moments passed as the lodge remained bathed in the eerie white light.

Just as quickly as they had dispersed, the swirling clouds of the fierce blizzard coalesced once more, and the white glow of the moon disappeared behind their thick, dark mass. The soft light had ended and the wind whipped across the wilderness once more. The torrent of snow pelted the lodge yet again, blasting it with frigid natural power.

All of the room's occupants looked at each other in awe. The parting of the clouds was a strange event, and all of them knew instinctively that it was a mystical omen, a powerful portent. Seeing the tribals smiling at what had just happened, the locals knew that it more than just coincidence; the moon had significant importance to the youths.

Empowered by the natural event, Tani held his mug in front of Jared, who responded to the call. Smacking their mugs together, the tribals gave a toast.

"To Scarskin!" they both said in excited tones, slamming back the last of their beer.

For the rest of the evening, the tribals sat at the mighty table and listened to the locals spin tales about the wilderness beyond the warmth of the mighty lodge. In a dreamy haze they sat, images of the full moon still burning in their memories. The thirteenth moon was a powerful accomplishment for both Jared and Tani, an event that distinguished them as survivors. Only seven more moons needed to rise before the tribals could return home, a tangible goal that inspired them and lifted their hearts. Though the road forward would not be easy, they were both one step closer to home.

Chapter 47
The Vanishing

Smiling, Tani grasped the wire cutters in his stumpy fingers. Working methodically, he cut the wiring in several places. He grasped a primitive soldering iron and pressed the solder to the connections, focusing intently on the task at hand. Smoke rose while the hot tip of the soldering iron melted the soft metal. Within mere seconds, the basic circuit board was complete. Elated by his success, Tani jumped up from his chair and rushed across his room.

Halfway across the room, the tribal genius stumbled over a tall pile of books. Determined to protect the newly created circuit board, Tani fell in such a way as to avoid damaging the precious device. Landing awkwardly on his knees, he managed to keep the circuit board intact. He ignored the pain from the fall and rose immediately, stepping over his beloved pile of books.

Rushing through his room once more, Tani reached his desk. A bundle of dynamite was lying out in the open. Reaching the hazardous explosives, he held his breath and crouched down. His breath stilled for a few lingering seconds as his green eyes pondered the device. He knew that his current circumstances were highly dangerous. If he made one wrong move attaching the circuit board to the explosives, they would detonate, killing himself and destroying his parents' house in the process.

Thinking the situation through, he nodded to himself. The tribal genius knew what needed to be done. Grasping the circuit board and the bundle of dynamite, Tani made the connections carefully. As he labored, a smile graced his lips. Many moments passed as he worked diligently, thinking only of the task at hand.

Concluding his work on the device, he sighed with a mixture of relief and accomplishment. He had completed his task and lived to tell the tale. Stretching, Tani walked across his room and collected the books that had been knocked to the floor in his flurry of excitement. Gently, he grasped each of them in turn, lovingly arranging them in the pile once more. As his hand brushed across each one, a jolt of excitement filled him. Such knowledge and power were housed in the ancient textbooks...

The dream of his home, his former home, passed into memory. Tani was not in his room in Scarskin; instead, he was in a much different place, a place far from home, far from safety. Instead of a warm room lit with soft light, Tani found himself in the freezing cold, staring into the night sky, watching the twinkling of distant stars. Instead of being in the safety of his former home, Tani found himself in the lonely wilderness pacing circles around the camp at night, standing guard over his companions. Lately, the tribal found his thoughts returning again and again to his former home, his former way of life. In times of panic and strife, he imagined what was most important to him: the pursuit of scholarly goals and a stable home. Thinking of his former home had given him solace in recent times.

More and more, he thought of his home. More and more, he missed his family and former way of life. The many months of travel had made the tribal youth homesick.

Sighing in the darkness, Tani eyed his eerie surroundings in awe. The moon, only a few days past full, was bathing the icy wilderness in a brilliant white light. A milky glow was reflecting off the newly fallen snow, bathing the wilderness in a serene ashen glow. Sweeping his eyes from side to side, he viewed the snow with calm glances. As he gazed around the wintry wilderness, Tani thought about the events that had led him to this very position, alone in the darkness, walking through the snow.

After the blizzard that had driven them into the lodge had subsided, Nova 7, under Lavosi's direction, had struck out once again into the wilds of the Darken Realm. After two days travel from the hunting outpost, the companions had ascended a steep mountain pass and found themselves in a secluded valley. The team had made camp as night settled over the forest. After eating a light meal, the members of Nova 7 settled into a chilling night's sleep out in the wilds. Tani was on patrol that night, having taken the first watch.

As he trudged through the snow in the soft light, an unusual sight greeted his eyes. Rushing forward, he came across a set of tracks. Alarmed, he crouched down and viewed them in concern. Whatever had left of the tracks walked on two legs and the tracks originated from Nova 7's camp. Inspecting the tracks, Tani was convinced that the trail had been made by none other than the vile Lavosi.

Seeing the tracks in the snow filled Tani with many questions. Where was Lavosi going? Why would he stray so far from camp, especially at night? Was the evil rat watching him, staring at Tani from the darkness?

His green eyes searched his surroundings with growing urgency. The trees were moving back and forth as a chill wind gripped the forest, blasting the foliage with a strong breeze. Expecting to see the sinister vermin lord, Tani was stunned by his disappearance. An eerie feeling settled over him as he scanned the dark forest for their guide.

The concern rapidly turned to anxiety. Trembling, Tani gripped his submachine gun with a shaky hand. Feeling increasingly wary, his half frozen hand pulled back the slide on his gun, chambering a fresh bullet into his weapon. He squinted once more in the darkness, his eyes scanning back and forth in search of the vermin king, his weapon ready. Lavosi had been creepy over the past months of travel, and Tani was afraid to confront him alone in the dark.

His anxiety mounting, the scholar was now searching for their guide with a frantic pace. Circling the entire camp, he still couldn't locate Lavosi.

"Damn it," he whispered under his breath. As he peered into the dark forest, the only thing Tani could see were the snow-encrusted trees waving back and forth in the cold wind. Lavosi was nowhere to be seen.

Turning round and round, he scanned the tree line. He half

expected to see Lavosi crouched in the darkness watching him, and was growing more anxious by the moment as the imagined encounter failed to materialize.

Grabbing his flashlight, Tani flipped the switch and pointed it into the snow. A line of tracks, clearly not human, could be seen heading north into the dark forest. Looking back at camp, Tani shook his head in frustration, then turned his attention back northwards, letting out a loud sigh. The smart thing to do would be to wake up Banion and the others. But if Lavosi was just messing around, Tani would feel really foolish getting everyone out of their warm sleeping bags to go search for the vermin king for no good reason. If he was just answering the call of nature, Lavosi would be on his way back to camp very soon. Not wanting to agitate the other members of Nova 7, Tani decided to investigate the strange disappearance on his own.

Gulping down his fear, he let his shaky hand lead the way, gripping the flashlight. As Tani pressed on, a burst of wind broke over the trees in the forest. Gently, the moving trees began to play tricks on Tani's fragile mind. Each waving branch seemed to materialize into an unseen horror. On several occasions, discolorations near the trees made Tani jump in fright. What seemed to be a bear moving near the trees quickly became nothing more than branches covered in snow. Illusion and icy mirage had gripped his psyche as he continued on his expedition into the frozen winter night. As he blinked several times, other images played on Tani's fears. A clawed arm swung forward and brushed across his face. Wanting to scream, Tani sighed in relief as he gripped a branch from a nearby pine tree. The wind had blown it into him as he passed near the foliage, startling him.

Steadying himself, the scholar moved deeper into the forest, further from safety, closer towards uncertainty. The tracks meandered back and forth, skirting the tree line.

It seemed like an hour to Tani, but a mere ten minutes passed as he followed Lavosi's tracks in the newly fallen snow. Cutting through an open meadow, he felt as if he were the only thing alive in the entire valley. The barren white expanse stretching around him as he traveled made him feel alone and totally vulnerable, exposed and defenseless. Feeling as if the entire world was watching him, Tani looked from side to side, expecting *something* to be tracking his advance. Much to his foolish dismay, nothing was watching him traverse the frozen meadow.

After he reached the far end of the open expanse of land, the tracks stopped abruptly at a rock formation. Unable to follow the trail any further, Tani was dumbfounded. Why would Lavosi travel so far away from their camp? Why would he leave the safety of the camp in the middle of the night?

Concern was quickly turning to suspicion. Still thinking that he might see the mutant rat totter down the rocks toward camp, Tani scanned his surroundings for many moments. As he waited, a chilling blast of wind rushed through the valley. Shuddering, the scholar abandoned the eerie scene. With confusion and urgency guiding him, Tani gave up on the search for Lavosi, deciding to return to camp and alert the rest of the team. Rushing full speed through the snow, Tani reached their camp quickly.

"Banion!" Tani called out in an excited tone.

The gunfighter was on his feet in flash. Having drawn his weapon, he grabbed Tani forcefully and stared him down with a feral look in his eyes. As he held his silver revolver close to the tribal's chest, Tani was terrified that he was about to get shot.

"Hold on!" he shouted again. "It's me!"

Having heard the commotion, Mineera and Jared were also on their feet.

"What's going on?" Jared asked, seeing the startled Banion still grasping Tani.

Mineera moved forward and grabbed Banion's arm, pulling him away from the tribal. "Tani, what is all of this about?"

"He's gone!" Tani declared in a near-stutter.

"Gone?" Banion echoed, holstering his weapon while rubbing his face with his left hand. "What the hell are you talking about?"

"Lavosi is gone!" Tani clarified.

After hearing his tale, they all scanned the camp. Lavosi's sleeping bag was still near the fire, but it was empty. He had left some of his other belongings in his wake, as well.

Blinking in disbelief, the rest of the team was confused. Tani was particularly distraught.

"Slow down," Mineera coaxed the tribal. "Tell us what happened again."

Taking a deep breath, Tani retold the tale of his discovery. When his story ended, a tremor of shock filled them. Where once there were five, now there were four. The sinister taint of Lavosi had ended. A dark presence that had infected the team was now missing.

As the team huddled anxiously, with more questions than answers, Banion crouched down and examined Lavosi's sleeping bag, grabbing a piece of parchment which had been left inside it. Shaking his head in rage, Banion clutched the paper fiercely. "Damn you!" he shouted out vehemently. "Damn you, Lavosi!"

"What is it, Banion?" Tani inquired.

Shaking his head, he held the parchment in the air. "Bastard left us a map."

"A map to what?" Jared quizzed.

"The military base," Mineera concluded correctly with an ominous tone. Her gift of intuition had answered the riddle.

The quest to locate a nuclear weapon was coming to a close. Having traveled a thousand miles with them, their guide had now mysteriously disappeared, leaving them a map to their prize.

Grabbing Tani's flashlight, Banion viewed the map intently, shaking his head in disbelief and growing frustration. "I can't believe this!"

"What?" Tani responded.

"According to this map, we are less than five miles away from the military base! That bastard led us into this valley and didn't even warn us. With Reaper Kai forces swarming the area, we are damn lucky they didn't find us and kill us. We were in harm's way and never even knew it!"

Hearing that they were a mere five miles from their objective, an objective that had been the focus of their lives for many months, each of the companions felt a fire light in their hearts. The end of the crusade was very near, so near that a few hours of travel would lead them to their goal. As the strange events of the night played in their minds, Nova 7 braced for the future. They had the information they needed to fulfill their quest, but it was still mired in danger. With Reaper Kai forces winning the great race and Lavosi's mysterious disappearance, Banion's team had to be on their guard.

The snowcapped peaks to the north were overwhelmed by a

surging mass of dark clouds. A fresh wash of white flakes fell from the sky with wild intensity. Feeling the chill night and the coming storm, Nova 7 converged around their campfire as another blizzard pelted the land.

Huddling together, they looked at each other with a fire rising in their hearts. Even though the cold wind blasted them, they stood resolute, fighting back the cold with the sheer power of will.

"We have two choices. That blizzard is about to hit. We either stay here and freeze our asses off, or break camp and assault the military base before the sun rises," Banion declared, eyeing each of them with a manic look in his eye.

"What about the Reaper Kai and Biogtechs? The enemy has over a hundred Biogtechs led by at least a dozen priests. What are we gonna do about them?" Tani quizzed.

Banion shook his head indifferently, seemingly unconcerned about the future. Grabbing his revolver, he checked it quickly, making sure it was loaded. Ignoring the tribal's concern, he simply smiled. "What's it gonna be? Freeze our asses off or end this bloody road once and for all?"

With overwhelming anxiety filling them, the reckless band knew what must be done. They didn't know how they would take on an army of enemy troops. It didn't matter; it simply didn't matter. Their nerves were shot, their confidence fading; the future was a blind curve rushing toward them. Though the coming strife was sure to be grim, each member of Nova 7 uttered a silent prayer, refusing to surrender, refusing to submit. Hope is an emotion devoted to a stubborn sense of justice; an emotion which is clung to so fiercely that death itself cannot abate it. To hell with fear, to hell with indecision. The world is led by those who refuse mediocrity and maintain an unwavering courage against tyranny.

"Are you with me?" Banion asked harshly.

"Damn right we're with you!" Jared replied with a grin.

"Let's march in there and take what's ours!" Tani shouted.

The trio looked at Mineera, who returned their gazes with a subtle, cryptic expression. Smiling, she spoke in a confident tone. "Destiny beckons; let's make history."

And so it was that a reckless band of heroes charged forth into the fray, against insurmountable odds, without a hint of self preservation. The future is for those who dare to dream, dare to defy the world, battle against inner fear, and hold a noble sense of unwavering justice against all odds.

   With hearts pounding, Nova 7 hastily grabbed their gear and charged towards the military base holding the holy grail of all ancient artifacts. Destiny was beckoning, and Nova 7 rushed forward with a fervent fury to meet it.

More information about the Darken Realm
can be found at
www.darkenrealm.com